Lecture Notes in Business Information Processing 276

Series Editors

Wil M.P. van der Aalst
Eindhoven Technical University, Eindhoven, The Netherlands
John Mylopoulos
University of Trento, Trento, Italy
Michael Rosemann
Queensland University of Technology, Brisbane, QLD, Australia
Michael J. Shaw
University of Illinois, Urbana-Champaign, IL, USA
Clemens Szyperski
Microsoft Research, Redmond, WA, USA

T0213737

More information about this series at http://www.springer.com/series/7911

Stefan Feuerriegel · Dirk Neumann (Eds.)

Enterprise Applications, Markets and Services in the Finance Industry

8th International Workshop, FinanceCom 2016
Frankfurt, Germany, December 8, 2016
Revised Papers

Editors
Stefan Feuerriegel
University of Freiburg
Freiburg
Germany

Dirk Neumann
University of Freiburg
Freiburg
Germany

ISSN 1865-1348   ISSN 1865-1356 (electronic)
Lecture Notes in Business Information Processing
ISBN 978-3-319-52763-5   ISBN 978-3-319-52764-2   (eBook)
DOI 10.1007/978-3-319-52764-2

Library of Congress Control Number: 2017930148

Printed on acid-free paper

This Springer imprint is published by Springer Nature
The registered company is Springer International Publishing AG
The registered company address is: Gewerbestrasse 11, 6330 Cham, Switzerland

Preface

Advancements in information and communications technology have paved the way for new business models, markets, networks, services, and players in the financial services industry. In the interest of further research advancements in this field of study, we organized FinanceCom 2016 in Frankfurt to help us understand, develop, and utilize the underlying systems, technologies, challenges, and opportunities. We invited leading academics from a broad range of disciplines, including computer science, business studies, media technology, and behavioral science, to discuss recent advances in their respective fields. This workshop also welcomed cross-disciplinary research work stemming from different backgrounds.

FinanceCom is part of an extremely successful workshop series that has taken place in different locations throughout the world, such as Regensburg (2005), Montreal (2007), Paris (2008), Frankfurt (2010), Barcelona (2012), and Sydney (2014). Its recent proceedings have been published as part of Springer's *Lecture Notes in Business Information Processing (LNBIP)* in order to reach a widespread audience. All papers accepted for inclusion have undergone a double-blind peer review process with subsequent mandatory revisions.

The 2016 workshop in Frankfurt was especially devoted to "The Analytics Revolution in Finance." Hence, submissions from the following areas of research were particularly welcomed:

Networks and business models

- Technology-driven transformation of the financial industry: - towards banking value networks
- Business process outsourcing/offshoring and information systems
- New e-finance business models enabled by IT
- New bank business models and challenges in a post-financial crisis
- Approaches to evaluating operational and credit risks as well as banking and market performance

Financial markets

- Electronic markets design and engineering
- Algorithmic and high-frequency trading/post-trading systems and infrastructures
- Analysis of intraday market data and news
- Regulation of electronic financial markets (e.g., MiFiD, EMIR or Dodd-Frank)
- Private equity and venture capital investments

IT and implementations

- Role of new technologies (e.g., Web services, cloud, big data, and grid computing)
- Implementation experiences and case studies
- Enabling decision support systems in banking and financial markets

- Enterprise communication in banking and financial services
- Interoperability of heterogeneous financial systems and evolving international standards

"New" emerging digital and virtual financial markets

- Virtual currencies (Bitcoin, Amazon, etc.)
- Alternative banking, loan, and financial market models
- New customer contact trends
- Crowdfunding, crowdsourcing, and B2B/B2C social media
- Loyalty card and smart card markets
- New banking and payment trends
- Banking, payments, and currencies in emerging countries

Conference theme: "The Analytics Revolution in Finance"

- Algorithms for automated and high-frequency trading
- Novel analytics approaches to risk modeling, e.g., Bayesian learning
- Utilizing big data for applications in finance
- Machine learning to support decision-making in financial markets
- New methodological approaches to deriving empirical results in finance research

Here, we provide a brief overview of the accepted publications.

- In their work "News Sentiment Impact Analysis (NSIA) Framework," Islam Qudah and Fethi A. Rabhi develop a system for news analytics activities. Their underlying goal is to quantify the impact of news sentiment from an arbitrary domain on the stock market. This work presents a combined approach that covers models, processes, and a corresponding software architecture.
- Ali Behnaz, Aarthi Natarajan, Fethi A. Rabhi, and Maurice Peat develop a semantic ontology for statistical learning. Their paper, "A Semantic-Based Analytics Architecture and Its Application to Commodity Pricing," demonstrates the ontology's capabilities in a case study in commodity pricing. The work thereby contributes to the standardization and model-driven work flow in data science.
- Qudamah Quboa, Brahim Saadouni, Azar Shahgholian, and Nikolay Mehandjiev suggest a path toward increasing underwriter profitability as part of their paper "Detecting Underwriters Stabilization Trades: A Clinical Study." In their study, the authors investigate the stabilization of shares in two large stock exchanges with the help of high-frequency tick data. Their empirical results provide an estimate of the profit from those trades.
- Petr Hajek, Vladimir Olej, and Ondrej Prochazka present the manuscript entitled "Predicting Corporate Credit Ratings Using Content Analysis of Annual Reports – A Naïve Bayesian Network Approach," wherein the authors utilize the financial statements of corporations in order to predict their credit ratings. For this purpose, they use naïve Bayesian network and latent semantic analysis in order to signal a low credit rating.

- Dorina Palade, Simon Alfano, and Dirk Neumann investigate the timing of corporate disclosures in their paper "Say It at the Right Time: Publication Time of Financial News." Since companies have the freedom to schedule the release of disclosures, their timing can provide valuable information regarding the content of the message and subsequent stock market returns.
- Liudmila Zavolokina, Mateusz Dolata, and Gerhard Schwabe prepared the manuscript on "FinTech Transformation: How IT-Enabled Innovations Shape the Financial Sector." This publication investigates the FinTech phenomenon from the perspective of information technology based on a collection of Swiss companies. Their results provide insights into the nature of FinTech innovations and outline the need for future research in this field of study.
- Florian Förschler and Simon Alfano examine the predictive relationship between financial news and the stock market as part of their work "Reading Between the Lines: The Effect of Language on Economic Indicators." For this purpose, they create a sentiment index based on ad hoc announcements and measure the directional influence based on Granger causality tests.
- Niklas Arvidsson, Jonas Hedman, and Björn Segendorf elaborate on "Cashless Society: When Will Merchants Stop Accepting Cash in Sweden—A Research Model." They suggest a research approach by which to study why shops accept or reject cash in Sweden, where it is left to the shop owners to choose which forms of payment they will accept. This can help to better understand potential shifts toward cashless payments in the future.
- Erika Matsak's "Credit Scoring and the Creation of a Generic Predictive Model Using Countries' Similarities Based on European Values Study" presents a data science approach to classifying credit scores. Here, the author studies transnational similarities and describes the benefits of using the generic predictive model in practice.

We would like to thank Peter Gomber and Florian Glaser (both of the University of Frankfurt, Germany) as part of the local organization team. In addition, we are grateful for the extraordinary work of all reviewers, authors, as well as the Program Committee.

If you are interested in joining this community centered around the study of financial markets, please feel free to join our mailing list or browse through the workshop website. To post to the e-mail list, please use the following address: finance-com@ambientmediaassociation.org; if you would like to subscribe to the e-mail list, please visit the following website: http://mail.ambientmediaassociation.org/mailman/listinfo/financecom_ambientmediaassociation.org.

November 2016

Stefan Feuerriegel
Dirk Neumann

Organization

Program Committee Chairs

Stefan Feuerriegel	University of Freiburg, Germany
Dirk Neumann	University of Freiburg, Germany

Organizing Committee

Stefan Feuerriegel	University of Freiburg, Germany
Florian Glaser	University of Frankfurt, Germany
Peter Gomber	University of Frankfurt, Germany
Dirk Neumann	University of Freiburg, Germany

Steering Committee

Marc Adam	University of Newcastle, Australia
Rajola Federico	Catholic University of the Sacred Heart, Italy
Stefan Feuerriegel	University of Freiburg, Germany
Peter Gomber	University of Frankfurt, Germany
Artur Lugmayr	Curtin University, Australia
Stefan Lessmann	Humboldt University of Berlin, Germany
Nikolay Mehandjiev	University of Manchester, UK
Jan Muntermann	University of Göttingen, Germany
Dirk Neumann	University of Freiburg, Germany
Maurice Peat	University of Sydney, Australia
Helmut Prendinger	National Institute of Informatics, Japan
Fethi Rabhi	University of New South Wales, Australia
Ryan Riordan	University of Ontario, Canada
Michael Siering	University of Frankfurt, Germany
Andrea Signori	Università Cattolica del Sacro Cuore, Italy
Christof Weinhardt	Karlsruhe Institute of Technology, Germany
Axel Winkelmann	University of Würzburg, Germany

Contents

News Sentiment Impact Analysis (NSIA) Framework

Islam Qudah[✉] and Fethi A. Rabhi

School of Computer Science and Engineering, University of New South Wales,
Sydney 2052, Australia
{ialqudah, fethir}@cse.unsw.edu.au

Abstract. News analysis activities have been the focus of many research studies across various life domains. So often, the goal of these studies is to automatically, analyze the meaning of news, and to gauge their impact on a particular domain. In this paper, we focus on studying sentiment analysis impact, on financial markets. Current studies, lack systematic approaches to evaluate the impact of a given sentiment dataset, in different financial contexts. We introduce a framework that encompasses models, processes, and a supporting software architecture for defining different financial contexts and conducting sentiment data-sets evaluation. The paper, describes a prototype implementation of the framework and a case study, which investigates the efficacy of the framework in evaluating the impact of a particular news sentiment dataset. The results demonstrate the capability of the framework in bridging the gap between producing a sentiment dataset and evaluating its impact in various financial contexts.

1 Introduction

News analysis activities have been the focus of many research studies across various life domains. So often, the goal of these studies is firstly, to automatically grasp the meaning of the news, and secondly to gauge their impact in a particular domain. To achieve the goal, various news analysis techniques were tested, ranging from simple to more sophisticated techniques. Among these techniques is text mining [11, 15] and sentiment analysis or opinion mining, which produces sentiment scores testifying the strength of the sentiment expressed in the text [12]. Our research project focuses on studying sentiment analysis in the context of financial markets. Reviewing the literature reveals to us a double facet problem. From one angle, we found inconsistencies in conducting impact analysis activities of any given sentiment dataset. Users follow different processes and routines to conduct impact analysis studies, and so often, the set of tools and applications used, are also different. Thus, clear and concise steps are needed, to enable users to conduct impact analysis activities in systematic way. From another angle, the existing impact analysis studies, evaluate their solutions in a fixed context setting. Varying the setting, could imply different results. Overall, this leads to the following two research questions. The first one is how to model an impact analysis framework, where users can conduct evaluation of any sentiment dataset. The second one, is how to enable studying impact analysis in various financial contexts.

© Springer International Publishing AG 2017
S. Feuerriegel and D. Neumann (Eds.): FinanceCom 2016, LNBIP 276, pp. 1–16, 2017.
DOI: 10.1007/978-3-319-52764-2_1

The paper is structured as follows. The next section gives some background on sentiment analysis and presents various approaches for measuring the impact on financial markets. Section 3, spells out the proposed News Sentiment Analysis Framework, designed for the purpose of carrying out impact analysis evaluation of any given sentiment dataset. Section 4, describes a prototype of the framework and its application to a case study. Finally, Sect. 5 concludes this paper.

2 Background and Related Work

Researchers on sentiment analysis and its impact on financial markets have primarily approached this area from two perspectives. The first group of researchers usually focus on aspects related to financial markets modelling and evaluation, while the second one centers the majority of their efforts in crafting and validating new sentiment analysis techniques, and less attention paid to financial markets impact models.

Surveying literature in the first group of studies shows the majority of the researchers have produced sentiment scores by analyzing existing news sources (e.g. EDGAR [24, 25]) and measuring sentiment in news corpus using a range of sentiment analysis techniques. Most sentiment datasets were based mainly on using existing lexicons [8, 11, 13], while [2, 6] sourced sentiment datasets from third parties such as Thomson Reuters [26]. Perhaps, the main challenge this group of researchers faced was the technical challenge, i.e. the lack of technical computing knowledge have constrained them in the range of computing techniques they can employ to determine sentiment. This group investigated the impact of news sentiment scores on financial markets by applying different statistical analysis techniques (e.g. regression analysis tests) and in some cases, simulating trading strategies. The majority of the sampled studies have used daily stock returns, but other measures such as volatility, future earnings, and/or trading volumes have also been used. For instance, Tetlock et al. in [25] applied over 20 regression tests, factoring not only sentiment index, but a range of other variables to understand the impact on stock returns, firm earnings and trading volume. In general, understanding the behavior of financial markets is far more complicated and some researchers have explained the underlying complexity [3, 15].

In the second group, the focus was chiefly bent towards proposing innovative sentiment analysis techniques. These techniques were often rendered into the development of a prototype, which encapsulates processes to automatically analyze sentiment for specific text source. These studies, often engage a range of text mining approaches, or natural language processing techniques to analyze text and construct sentiment indices, such as [4, 8, 14, 20–22, 27]. These studies also applied regression analysis and in some cases trading strategies to measure the impact computed sentiment indices had on financial markets. However, these activities were limited and lacked intricate financial modelling knowledge found in the first group. Therefore, limited impact analysis is carried out. For instance, Schumaker et al. in [20] applied limited statistical significance tests to predict stock returns.

In summary, all the studies share some of the analysis processes and differ on some others. The first group used simple techniques, principally based on lexicons. the second group, devoted much more time to trial machine learning, and natural language processing approaches. On the other hand, the first group conducted comprehensive regression analysis tests, whereas limited impact analysis activities have been carried out by the first group. All studies lack a systematic approach, that assimilates the role of financial context, in evaluating the effectiveness of a given sentiment dataset. Researchers who wish to reuse part, or all of a particular evaluation process in a different financial context, or for a different news corpus hit a roadblock. The literature shows there is a gap in defining systematic and reusable evaluation processes that could be used by a wide range of users, to automatically conduct impact analysis of sentiment datasets in different financial contexts. To address this gap, this paper introduces a framework called News Impact Analysis (NSIA) framework, devoted to enable users to transparently conduct impact studies, where systematic approaches are presented, preserved and results are reproducible.

3 News Sentiment Impact Analysis (NSIA) Framework

In this section, we address the research gap identified in Sect. 2, by proposing a model, which consists of three components. These components are: the financial context in which the impact analysis is being conducted, the sentiment data selection criteria that been used in filtering the news, and the financial market impact measure used as indicator of reaction to news sentiment. Almost all studies reviewed have incorporated these components informally. The proposed model formally represents these components, and formalizes set of processes to guarantee impact analysis studies are conducted in a systematic and consistent way. The proposed *News Sentiment Impact Analysis (NSIA)* framework consists of three main components: NSIA Comparison Parameters Model (CPM), NSIA software architecture to support the model implementation and a set of NSIA processes to guide users in conducting evaluation studies.

3.1 Defining Comparison Parameters Model (CPM) Parameters

The Comparison Parameters Model (CPM) divides the contextual parameters into three sets: Financial market context parameters (C), Sentiment extraction parameters (SN) and Impact Measure parameter (IM). In Table 1, C parameters are defined as a context vector where $C = (E, Ev, B, Bv, P)$.

In Table 2, sentiment extraction parameters, defined by sentiment vector SN, where $SN = (X, FA, Tn)$.

Finally, the IM parameters represent the impact magnitude, of the news sentiment scores for a given set of C parameters as shown in Table 3. An earlier version of CPM parameters can be found in [18].

3.2 NSIA Architecture

The NSIA architecture is designed to support the evaluation of a sentiment dataset as defined by the CPM parameters. The NSIA architectural design follows the ADAGE

Table 1. Financial context parameters (*C*)

Parameter name	Definition	Example
Entity *E*	Entity being impacted by the news	Company, an industry sector, the economy of a country as a whole.
Entity Variable *Ev*	Variable associated with the entity in question whose value is impacted	Closing share price, an index, GDP etc.
Benchmark *B*	Benchmark against which the impact will be measured	List of companies, an industry sector, the economy of a country as a whole.
Benchmark Variable *Bv*	Value indicative of the selected benchmark	Closing share price, an index, GDP etc.
Study Period *P*	The period during which the evaluation takes place	Days, Months, years…etc.

Table 2. Sentiment extraction parameters (*SN*)

Parameter name	Definition	Example
Sentiment Dataset *M*	Name of sentiment dataset being evaluated	AlchemyAPI [1], Semantria Lexalytics [10], Quandl [17] and RavenPack [19]
Filtration functions *FA*	Selecting records of interest based on the attributes (fields) of a news sentiment record denoted as $\{a1, a2\ldots an\}$	(Sentiment class = positive), (sentiment score > 0), (news type = Alert) …etc.
Extreme Sentiment Extraction (*ESE*) Algorithms	Ranking algorithms, that define the basis for selecting "extreme" news sentiments	For example, extract top 5% negative news records

Table 3. Impact measure parameters (*IM*)

Parameter name	Definition	Example
Impact Measure Parameter *IM*	Specifies how to measure the impact of news sentiment on the entity (*E*), relative to the benchmark (*B*)	Mean Accumulative Abnormal Returns, Intraday returns, trading volume, market depth …etc.
Estimation Period *EP*	The estimation period used to measure impact	Hours, Days, Months..etc.

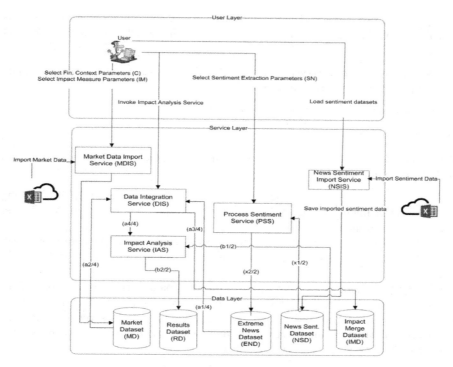

Fig. 1. NSIA architecture overview

Framework guidelines [7, 9]. It is a service oriented architecture, which encompasses three layers: User layer, Service layer and Data layer as illustrated in Fig. 1. The User layer mediates the interactions between users and the Service layer, based on user selections. The Service layer consists of, number of web services, which encapsulate majority of the framework's business logic, these services can be described as follows:

1. Market Data Import Service (MDIS): based on user selection of CPM parameters, via the user interface, this service imports market data like, stock prices from heterogeneous data sources (web based, excel, flat files) and commit them to the Market Dataset (MD) component of the Data Layer.
2. News Sentiment Import Service (NSIS): similar to MDIS, this service imports a sentiment dataset according to relevant CPM parameters from a range of sources, and loads them into the News Sentiment Dataset (NSD).
3. Process Sentiment Service (PSS): based on CPM parameters, this service reads the NSD, applying the user selection parameters and generates a subset of NSD called Extreme News Dataset (END).
4. Data Integration Service (DIS): merges extreme news and market data into Impact Measures Dataset (IMD).

5. Impact Analysis Service (IAS): responsible of executing impact analysis models defined in CPM, reading IMD and generating results, which get saved into Results Dataset (RD).

The Data Layer comprises five datasets which preserve all impact studies related data from raw data to processed data.

3.3 NSIA Processes

NSIA includes a set of well-defined processes that are designed to guide users to conduct evaluation studies:

- Loading market data: this process reads the CPM parameters then invokes the Market Data Import Service (MDIS), which is responsible of loading the market data related to the context (C).
- Loading news sentiment datasets: this process reads the CPM parameters then invokes News Sentiment Import Service (NSIS) to import any given sentiment dataset.
- Identifying extreme news records: based on CPM parameters, this process invokes Process Sentiment Service (PSS), which processes NSD and produce END.
- Conducting impact analysis: based on CPM parameters, this process performs impact analysis using the Impact Analysis Service (IAS), Fig. 2 shows an example of impact analysis process sequence diagram.

As we are using Service Oriented Architecture (SOA), all these processes can be automated using Business Process Modelling (BPM) or workflow technologies.

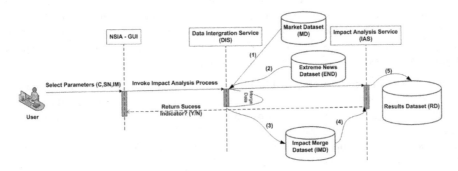

Fig. 2. NSIA impact analysis sequence diagram

4   Case Study

4.1   NSIA Prototype Implementation

To enable testing the usability and functionality of NSIA framework, we have developed a prototype, which implements the components described earlier in Sect. 3.2 (NSIA architecture). The implementation has been carried out as follows:

- Market Data Import Service (MDIS) we have reused an existing implementation of this service available on the ADAGE framework [7]. This implementation loads market data from Thomson Reuters Tick History (TRTH) API into the Market Dataset (MD). Market data loaded is constrained by the Financial Context Parameters (C), defined in next section.
- News Sentiment Import Service (NSIS) used to import commercially published sentiment dataset in (csv format) into News Sentiment Dataset (NSD). The dataset used in the prototype is Thomson Reuters News Analytics (TRNA) [26]. The web service implemented using Java programming language, Web Service Definition Language (WSDL) and Simple Access Object Protocol (SOAP). NSIS imported TRNA into NSD dataset.
- Processing Sentiment Service (PSS) and Data Integration Service (DIS): were all built using the Java programming language and made accessible via a Web Service Definition Language (WSDL) interface and Simple Access Object Protocol (SOAP) calls.
- Impact Analysis Service (IAS): we have reused an existing implementation of this service available on the ADAGE framework that uses Eventus [5].

The Data Layer encompassing all the datasets has been implemented using Oracle 11 g database management system to preserve data related to the experiments conducted.

The first objective of the prototype is to evaluate CPM via two experiments, one to evaluate the financial context component (C) of the model, and the other is to evaluate the sentiment extraction component (SN) of the model. The Impact Measure parameter (IM) will be fixed to Daily Mean Accumulative Returns $MCAR$ across all the experiments. Estimation Period EP parameter will be set to (−30 days to +30 days) before and after an extreme event day. The second objective is to evaluate the user experience aspect of NSIA framework.

4.2   Setting Financial Context Parameters C

The initial CPM model introduced in [18] was tentatively evaluated using 2 stocks listed on the Australian financial market. The results of the study concluded that further testing is needed to understand the impact of sentiment datasets in various financial

Table 4. Financial contexts parameters

Ctx C	Entity E (Ev = Daily closing value)	Benchmark B (Bv = Daily closing value)	Benchmark Description	Study Period P
$C1.1$	{ATLI}[1] ∩ {AORD}	ATLI	ATLI = Australia's ASX top 20 leaders	(1/01/20110,31/12/2011)
$C1.12$	{ATLI} ∩ {AORD}	AORD	AORD = Australia's All Ordinaries Index	(1/01/20110,31/12/2011)
$C2.1$	{GDAXI} ∩ {CXKNX}	CXKNX	CXKNX = Germany's Industrial Index	(1/01/20110,31/12/2011)
$C2.2$	{GDAXI} ∩ {CXKNX}	GDAXI	GDAXI = Germany's DAX Index	(1/01/20110,31/12/2011)
$C3.1$	{GTSX} ∩ {SPTSE}	GTSX	GTSX = Canada's healthcare index	(1/01/20110,31/12/2011)
$C3.2$	{GTSX} ∩ {SPTSE}	SPTSE	SPTSE = S&P Toronto Stock Exchange	(1/01/20110,31/12/2011)
$C4.1$	{DJI} ∩ {HWI}	DJI	DJI = Dow Jones Industrial Average Index	(1/01/20110,31/12/2011)
$C14.2$	{DJI} ∩ {HWI}	HWI	HWI = NYSE Arca Computer HW index	(1/01/20110,31/12/2011)

[1]{ATLI} means all the constituents of market index ATLI, and likewise, all other market indices, Ev is the daily closing price for every constituent in entity E.

contexts. This case study broadens the financial context by selecting 377 stocks listed on 8 financial market indices, in 4 countries.

In this first experiment we define and test the CPM model in multiple financial contexts C settings. The set of financial context parameters are defined in Table 4.

4.3   Setting Sentiment Extraction Parameters (SN)

This section describes how sentiment extraction parameters are set. The goal is to understand how impact results are affected by the way we identify extreme negative news records provided by TRNA dataset. The sentiment extraction is conducted in two phases: Filtration Functions (FA) and then Extreme Sentiment Extraction (ESE) algorithms. The notation $\{n1, n2...na\}$ represent all attributes of a given sentiment dataset n. TRNA attributes and filtration values are defined in Table 5.

We define two extreme sentiment extraction algorithms ESE_{VOL} and ESE_{Tot}. ESE_{VOL} and ESE_{TOT}. We also define an additional algorithm (referred to as ALL_{NEWS}) which selects all news records of an entity E, this algorithm is used for benchmarking against ESE_{VOL} and ESE_{TOT}.

Table 5. TRNA attributes and filtration values

Attribute no.	TRNA attributes	FA filtration values	Description
1	RIC	RIC must be equal to Entity (E) found within Context (C)	Reuters Instrument Code (RIC), stores stock code as listed on the financial market index, for instance DAX Industrial index had 36 stocks listed between 2010 and 2011 study period. FA will filter news records related to entity E
2	RELEVANCE	1	News record must be of high relevance to the entity E in question, values range from 0 to 1.
3	NO_COMP	1	The news record talks about one single stock, which should be the Entity E related to the experiment. Values can be ≥ 1.
4	LNKD_CNT5	0	News record is novel, and never been reported in any other news stories, up to a week from the day of news story release.
5	SENT_CLASS	{-1,0,1}	Sentiment Class parameter, -1 for negative class, 0 for neutral and 1 for positive class.
6	SENT_SCORE	[0,1]	Sentiment Scores attribute, defines the weight of sentiment class. Value ranges from 0 to 1
7	StoryDate (dd/mm/yyyy)	[1/01/20110,31/12/2011]	Date when news took place.

The first algorithm ESE_{VOL}: which is a volume based algorithm, computes the ratio of number of positive news with number of negative news within a day, if the ratio below the threshold parameters, then the news record is tagged as extreme news record. Assuming a function count(n_a), which will return the number of news records in dataset n_{a_i}, this algorithm is described as follows:

```
ESE_VOL (Pr, Ps, NSD):
/* 1. Get NSDF Average and standard deviation of the whole
sentiment dataset NSD*/
/*AvgN stores the average count of negative tagged news across
the whole dataset*/
AvgN= μ(count (NSD) ) where nS_ENT_CLASS=-1
/*AvgP stores the average count of positive tagged news across
the whole dataset */
AvgP= μ(count (NSD) ) where nS_ENT_CLASS=+1
/*StdN stores the standard deviation of the count of negative
tagged news across the whole dataset*/
StdN = σ(count (NSD)) where nS_ENT_CLASS=-1

/*2. Now we loop through V and get daily counts of positive and
negative    tagged news records, which will be compared to
average counts of the whole set stored above*/

V= Select all days (d) in dataset NSD

For every day d in V do

/*   dailyNCount   stores   the   count   negative   tagged   news
records for day d */
    dailyNCount(d)  = Count (NSD)) where day = d and nS_ENT_CLASS=-1

/* dailyPCount stores the count negative tagged news records for
day d */
    dailyPCount(d)  = Count (NSD) where day = d and nS_ENT_CLASS=+1

/* Calculate the ratio of positive to negative news in day d*/
    RatioCount(d)  = ((dailyPCount(d) ×100)/dailyNCount(d))

/*If the daily ratio calculated above is less than Pr and daily
count of negative news is greater than AvgN + Ps * StdN then
this is a record we define as extreme */
 If (RatioCount(d) ≤ Pr and
    (dailyNCount(d)>(AvgN+(Ps× StdN)))) then
  add all records of NSD with day d to Subset (END)
 End if
End For
Return (END)
```

The second algorithm E_{TOT}, which takes into consideration sentiment scores, computes the ratio of the total sentiment scores of positive news with total sentiment scores of negative news within a day, if the ratio is below the threshold parameters, then the news record is tagged as an extreme news record. Assuming a function sum (n = SENT_CLASS) that will return the sum of sentiment scores, of attribute SENT_CLASS within a TRNA news record, this algorithm can be described as follows:

```
ESE_TOT(Pr, Ps, NSD)=
/*1. Get NSDF Average and standard deviation of the whole
sentiment dataset NSD*/

/*AvgN stores the mean of total of sentiment scores of negative
tagged news   for all records in dataset NSD */
AvgN=μ(sum (NSD.SENT_SCORE)) , where nSENT_CLASS=-1

/*StdN stores the standard deviation of total of sentiment
scores of negative tagged news for all records in dataset NSD */
StdN = σ(sum (NSDF.SENT_SCORE)) where nSENT_CLASS=-1

/*2. Now we loop through V and get daily total of sentiment
scores SENT_SCORE of negative and positive tagged news records,
which will be compared to means of the whole set stored above*/
V= Select all days (d) in dataset NSD
For every day d in V do

/* dailyNSum stores the sum of sentiment scores of negative
tagged news     records for day d */
dailyNSum(d)=  sum   (NSD.SENT_SCORE)   where   day  =  d   and
nSENT_CLASS=-1
/* dailyPSum stores the sum of sentiment scores of positive
tagged news     records for day d */
  dailyPSum(d)  =  sum   (NSD.SENT_SCORE)   where   day  =  d   and
nSENT_CLASS=+1

/* Calculate the ratio of positive to negative news in day d*/
  RatioTot(d) = ((dailyPSum(d) ×100)/(dailyNSum(d)))

/*If the daily ratio calculated above is less than Pr and daily
sum of negative news is greater than AvgN + Ps * StdN then this
is a record we define as extreme */
  If (RatioTot(d)≤ Pr and
     (dailyNSum(d)> (AvgN + (Ps× StdN))))) then
       add all records of NSD with day d to Subset (END)
  End if
End For
Return (END)
```

ESE_{VOL} and ESE_{TOT} both make use of user defined threshold parameters, to rank news records in dataset (see Table 6 for values used in our experiments).

4.4   Experiments and Results

As stated earlier, the objective of these experiments is evaluating two components of the Comparison Parameters Model (CPM), the first component is the financial context (C) and the second component is the sentiment extraction (SN) parameters. We only investigated the impact of *negative* tagged news in TRNA using Mean Accumulative Abnormal Returns (MCAR). We conducted 24 experiments; where each context parameters set (C) defined in Table 4 (Sect. 4.2), was tested against the extreme

Table 6. Extreme Sentiment Extraction (*ESE*) parameters

ESE (Pr, Ps, NSD)	ESE_{VOL}	ESE_{TOT}	Description
Pr	0.2	0.3	Threshold parameter to define the ratio between negative and positive news for a day.
Ps	0.5	0	Threshold parameter multiplied by the standard deviation to define how far we want the algorithm to deviate from the mean.

sentiment extraction algorithms defined in Sect. 4.3. The TRNA dataset used in this case study contains 4,359,099 distinct news items. The experiments examined news records related to 377 stocks, listed on 8 financial market indices, located in 4 countries {Australia, Canada, Germany, USA}. The experiment results are shown in Table 7.

Table 7. Experiments' Parameters and Impact Results

Exp. no.	Fin. context. (C)	Sentiment data SN			Impact results			
		$|NSD|$	ESE Algo.	$|END|$	MCAR	Precision weighted (CAAR)%	Patell Z	Generalized sign Z
1	c1.1	1333	ESE_VOL	30	−0.18%	−0.03%	−0.078	−0.126
2	c1.1	1333	ESE_TOT	28	−0.84%	−1.05%	−4.601[d]	−1.870[b]
3	c1.1	1333	ALL_NEWS	1333	−0.05%	−0.06%	−1.571[a]	−0.007
4	c1.2	1333	ESE_VOL	30	−0.25%	−0.12%	−0.381	0.265
5	c1.2	1333	ESE_TOT	28	−0.82%	−1.05%	−4.496[d]	−1.383[a]
6	c1.2	1333	ESE_VOL	1333	−0.04%	−0.06%	−1.635[a]	−0.284
7	c2.1	776	ESE_VOL	37	−0.39%	0.01%	0.027	0.200
8	c2.1	776	ESE_TOT	34	0.07%	0.13%	0.619	−0.661
9	c2.1	776	ALL_NEWS	776	0.16%	0.15%	2.792[c]	2.016[b]
10	c2.2	776	ESE_VOL	37	−0.31%	0.00%	−0.020	0.926
11	c2.2	776	ESE_TOT	34	0.28%	0.36%	1.066	0.622
12	c2.2	776	ALL_NEWS	776	0.11%	0.08%	2.254[b]	1.453[a]
13	c3.1	119	ESE_VOL	2	−0.84%	−0.84%	−0.711	−1.022
14	c3.1	119	ESE_TOT	10	−0.22%	−0.15%	−0.597	−0.473
15	c3.1	119	ALL_NEWS	119	0.02%	0.05%	0.223	0.315
16	c3.2	119	ESE_VOL	2	−1.23%	−1.23%	−0.753	−0.874
17	c3.2	119	ESE_TOT	10	−0.07%	−0.10%	−0.114	1.353[a]
18	c3.2	119	ALL_NEWS	119	−0.02%	−0.01%	−0.502	−0.291
19	c4.1	1183	ESE_VOL	28	−0.10%	−0.29%	−1.289[a]	0.575
20	c4.1	1183	ESE_TOT	15	0.69%	0.22%	0.507	−0.265
21	c4.1	1183	ALL_NEWS	1183	−0.02%	−0.03%	−0.991	1.833[b]
22	c4.2	1183	ESE_VOL	28	0.32%	0.30%	1.140	2.953[c]
23	c4.2	1183	ESE_TOT	15	0.95%	0.76%	1.649[b]	1.482[a]
24	c4.2	1183	ALL_NEWS	1183	−0.03%	−0.04%	−1.091	0.501

The symbols a, b, c, and d denote statistical significance at the 0.10, 0.05, 0.01 and 0.001 levels, respectively, using a generic one-tail test. |END| represents the number of extreme news records returned by ESE algorithms. |NSD| represents the number of distinct news items retrieved by *FA* function.

4.5   Discussion

The impact of negative sentiment scores is tested on the stock returns (MCAR column) shown in Table 7. Thus, filtering sentiment datasets with negative scores, should infer lower returns. The lower MCAR figures, the better the results and vice versa. NSIA framework implemented two ESE algorithms, we discuss how these algorithms performed, and in which financial context (C) they make more sense.

The results in Fig. 3 aggregates the MCAR figures by ESE algorithm. it shows that returns do in fact react news volume more than other filtering techniques. The results (experiments 1,4,7,10,13,16,19,22) show high drop of almost 3% of accumulated abnormal returns. Drilling down by country, to verify if news volume had the same effect across all the financial contexts. The results illustrated in Fig. 4, show that sentiment scores filtration technique, (ESE_{TOT}), performed better than news volume filtration technique (ESE_{VOL}) in the context of the Australian markets. Experiments 2

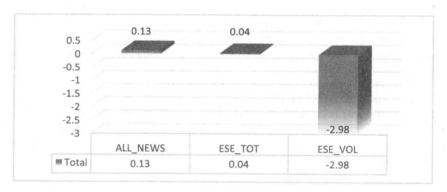

Fig. 3. Sum of all MCARs for the 24 experiments by ESE Algorithms

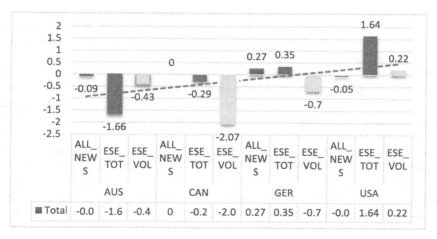

Fig. 4. Impact grouped by country and Extreme Sentiment Extraction Algorithms

and 5, show an accumulative drop of 1.66% below 0. The results disclose, that the Australian and Canadian markets were the best responsive to TRNA dataset, as compared to the US and German markets. The results demonstrated the varying effect filtration techniques, played in implying different impact results.

The news sentiment dataset TRNA impact results, entail news volume based trading strategy, works well in Canadian markets, and not recommended to other markets (Australian, American and German) markets. The Australian markets, seems to be more responsive to the subjectivity of the news, more than the volume of negative news. a conclusion we can make, when looking up the (ESE_{TOT}) MCAR figures for Australia. The benchmark algorithm ALL_{NEWS} (experiments 3,6,9,12,15,18,21,24), which naively selects all news records regardless of it is sentiment score, show MCAR figures have not been affected by this filtration technique.

The second aspect investigated in this case study was the user experience in running the impact studies. All the experiments, involved users had some background knowledge in conducting impact analysis studies. All the processes were manually executed by the users, and it took only few hours for the users to conduct the 24 impact analysis experiments. All the experiments are replicable, as the experiments' raw data, filtered data, and the sets of parameters used are preserved inside the prototype, and can be recalled at any time. All processes can be easily automated using workflow technology, such as TAVERNA [16], from import marketing data, importing sentiment data, to processing sentiment data and impact analysis.

The case study successfully demonstrated the functionality and capability aspects of the NSIA framework as well as opportunity for automation. The case study successfully demonstrated how it implemented a prototype, to test the efficacy of a commercially available sentiment dataset (TRNA), not only in one fixed setting, but rather in multiple financial context settings.

5 Conclusion and Future Work

This paper proposed a News Sentiment Impact Analysis (NSIA) framework to guide users in conducting evaluation studies of sentiment datasets. To enable testing the usability and functionality of NSIA framework, we have developed a prototype and validated its performance using a case study.

NSIA framework demonstrated a design methodology, that enables evaluating the impact of any given sentiment dataset in a systematic way. A set of contextual parameters representing the financial context chosen, the sentiment-related parameters and the impact measure parameter were together incorporated within the framework. To support the researchers in using the methodology, a service oriented architecture and associated prototype are proposed. The experiments, and the results we present in the case study are just a testimony and proof of concept, to demonstrated the functionality of the framework. The results of the case study, reveal to certain extent the importance of considering multiple financial contexts, when evaluating a given sentiment dataset. One main conclusion is that the choice of the daily MCAR as an impact measure does conceal a lot of volatilities that could impact the financial context during the day. Changing the impact measure to more news sensitive measures, such as, intraday returns, liquidity, and price jumps could vary the results significantly. So, this will be carried out as part of the next round of validation of NSIA framework. We are also in the process of incorporating more sentiment datasets, to validate the framework's ability in adapting to different types of news and sentiment measures.

The experiments also show some limitations of the prototype that could be improved in future work, such as improving the GUI to include visualizations of results. Automating the processes using workflow technology is another area of improvement as it would allow users to conduct thousands of experiments that correspond to different combinations of CPM parameters.

Acknowledgments. We are grateful to Sirca [23] and Thomson Reuters [26] for providing access to the data used in this research. We are also grateful to Brahim Saadouni from the Manchester Business School, for helping on different aspects of this research work.

References

1. AlchemyAPI (2016). http://blog.mashape.com/list-of-20-sentiment-analysis-apis/. Accessed June 2016
2. Allen, D.E., McAleer, M., Singh, A.K.: Daily market news sentiment and stock prices (2015)
3. Baker, M., Wurgler, J.: Investor sentiment and the cross-section of stock returns. J. Finan. **61**(4), 1645–1680 (2006)
4. Bollen, J., Mao, H.: Twitter mood as a stock market predictor. Computer **44**(10), 0091–94 (2011)
5. Cowan, A.R.: Eventus 8.0 User's Guide. Cowan Research LC, Ames (2007)
6. Dzielinski, M.: News sensitivity and the cross-section of stock returns (2011). www.nccr-finrisk.uzh.ch
7. Rabhi, F.A., Yao, L., Guabtni, A.: ADAGE: a framework for supporting user-driven ad-hoc data analysis processes. Computing **94**(6), 489–519 (2012). 10.1007/s00607-012-0193-0
8. Hagenau, M., Korczak, A., Neumann, D.: Buy on bad news, sell on good news: how insider trading analysis can benefit from textual analysis of corporate disclosures. In: Workshop on Information Systems and Economics (WISE 2012), Orlando, Florida, USA (2012)
9. Yao, L., Rabhi, F.A.: Building architectures for data-intensive science using the ADAGE framework. Concurrency Comput. Pract. Experience **27**(5), 1188–1206 (2015)
10. Lexalytics (2016). https://www.lexalytics.com/. Accessed June 2016
11. Li, F.: Do Stock Market Investors Understand the Downside Risk Sentiment of Corporate Annual Reports (2007)
12. Liu, B.: Sentiment analysis and opinion mining. Synth. Lect. Hum. Lang. Technol. **5**(1), 1–167 (2012)
13. Loughran, T., McDonald, B.: When is a liability not a liability? textual analysis, dictionaries, and 10-Ks. J. Finan. **66**(1), 35–65 (2011)
14. Lugmayr, A.: Predicting the future of investor sentiment with social media in stock exchange investments: a basic framework for the DAX performance index. In: Friedrichsen, M., Mühl-Benninghaus, W. (eds.) Handbook of Social Media Management. Media Business and Innovation, pp. 565–589. Springer, Heidelberg (2013)
15. Nassirtoussi, A.K., Aghabozorgi, S., Wah, T.Y., Ngo, D.C.L.: Text mining for market prediction: a systematic review. Expert Syst. Appl. **41**(16), 7653–7670 (2014)

16. Oinn, T., Addis, M., Ferris, J., Marvin, D., Senger, M., Greenwood, M., Carver, T., Glover, K., Pococ, M.R., Wipat, A., Li, P.: Taverna: a tool for the composition and enactment of bioinformatics workflows. Bioinformatics **20**(17), 3045–3054 (2004)
17. Quandl, Investor sentiment data (2016). https://www.quandl.com/data/AAII/AAII_SENTIMENT-AAII-Investor-Sentiment-Data. Accessed August 2016
18. Qudah, I., Rabhi, F., Peat, M.: A proposed framework for evaluating the effectiveness of financial news sentiment scoring datasets. Enterprise Applications and Services in the Finance Industry. Lecture Notes in Business Information Processing, vol. 217. Springer, Heidelberg (2014)
19. RavenPack, RavenPack news scores user guide RavenPack (2010)
20. Schumaker, R.P., Zhang, Y., Huang, C., Chen, H.: Evaluating sentiment in financial news articles. Decis. Support Syst. **53**(3), 458–464 (2012)
21. Siering, M.: "Boom" or "ruin"–does it make a difference? using text mining and sentiment analysis to support intraday investment decisions. In: 2012 45th Hawaii International Conference on System Science (HICSS), pp. 1050–1059 (2012)
22. Siering, M.: Investigating the impact of media sentiment and investor attention on financial markets. In: Rabhi, F.A., Gomber, P. (eds.) Enterprise Applications and Services in the Finance Industry. Lecture Notes in Business Information Processing, pp. 3–19. Springer, Heidelberg (2013)
23. Sirca (2016). http://www.sirca.org.au/. Accessed June 2016
24. Tetlock, P.C.: Giving content to investor sentiment: the role of media in the stock market. J. Finan. **62**(3), 1139–1168 (2007);
25. Tetlock, P.C., Saar-Tsechansky, M., Macskassy, S.: More than words: quantifying language to measure firms' fundamentals. J. Finan. **63**(3), 1437–1467 (2008)
26. Thomson Reuters, Thomson Reuters News Analytics (TRNA) (2010). http://thomsonreuters.com/products/financial-risk/01_255/news-analytics-product-brochure–oct-2010.pdf. Accessed July 2016
27. Vu, T., Chang, S., Ha, Q.T., Collier, N.: An experiment in integrating sentiment features for tech stock prediction in twitter (2012)

A Semantic-Based Analytics Architecture and Its Application to Commodity Pricing

Ali Behnaz[1]([⊠]), Aarthi Natarajan[1], Fethi A. Rabhi[1],
and Maurice Peat[2]

[1] School of Computer Science and Engineering,
University of New South Wales, Sydney, NSW 2052, Australia
{ali.behnaz, f.rabhi}@unsw.edu.au,
aarthi22@optusnet.com.au
[2] The University of Sydney Business School, Sydney, NSW 2006, Australia
maurice.peat@sydney.edu.au

Abstract. Over the past decade, several sophisticated analytic techniques such as machine learning, neural networks, and predictive modelling have evolved to enable scientists to derive insights from data. Data Science is characterised by a cycle of model selection, customization and testing, as scientists often do not know the exact goal or expected results beforehand. Existing research efforts which explore maximising automation, reproducibility and interoperability are quite mature and fail to address a third criterion, usability. The main contribution of this paper is to explore the development of more complex semantic data models linked with existing ontologies (e.g. FIBO) that enable the standardisation of data formats as well as meaning and interpretation of data in automated data analysis. A model-driven architecture with the reference model that capture statistical learning requirement is proposed together with a prototype based around a case study in commodity pricing.

Keywords: Ontologies · Semantic · Analytics · Commodity · Statistical learning · FIBO · Architecture · ADAGE · Model-driven engineering · Big data · Data science

1 Introduction

Many areas of science such as geo-sciences, astronomy, genomics and computational physics are confronted with the exponential growth of data. This data presents a vital opportunity to research scientists to understand the behaviour of complex systems and gain fundamental insight. The advent of e-commerce has produced similar growth in economic and business data e.g., security market data, sales forecasts, economic forecasts, inventory studies, workload projection, utility studies, budget analysis, etc. In this paper, we are particularly interested in the analysis of data which consists of observations measured sequentially at discrete points of time, commonly known as time series, e.g., interest rates and exchange rates recorded daily. This data is temporal in nature, it can be modelled deterministically with functions of time and analysed to extract meaningful information that help to better understand the dynamics and

© Springer International Publishing AG 2017
S. Feuerriegel and D. Neumann (Eds.): FinanceCom 2016, LNBIP 276, pp. 17–31, 2017.
DOI: 10.1007/978-3-319-52764-2_2

distribution of the data, e.g., draw statistical inferences from the observed data to guide decision making or make predictions about future values of the data based on the previously observed values (forecasting).

Regardless of scientific domain, data analysts are confronted by a number of challenges. Firstly, data analysis is a *complex, time-consuming process* requiring data analysts to combine several independent steps; accessing distributed data sources, local and remote custom software components (e.g. algorithms and scripts) and specialised tools (e.g., statistical tools, mathematical packages) into a larger analysis "pipeline" or process. At a high-level, this pipeline can be divided broadly into four phases: data acquisition, data preparation, data analysis and visualisation [27] (see Fig. 1).

Fig. 1. Different phases in the analytics pipeline

Over the past decade, sophisticated analytic tools and multidisciplinary techniques such as machine learning, neural networks, predictive modelling and data-mining have evolved to enable analysts to derive the needed insights from data [11]. Even though these techniques have emerged as popular strategies for complex analytics, they do not provide an overall solution to analysts conducting *in-sillico* data-intensive analysis. These techniques constitute one element of the overall analysis pipeline, analysts require a broader solution that captures all the phases of data analysis into a "integrated whole". The primary focus in most research efforts has been the creation of new theories, techniques and software to deal with the complex characteristics of data, practical analytic challenges as perceived by a domain-user have received little or no attention. As data-processing tools and applications are largely developed by software developers, often written in proprietary formats with competing specifications, standards and frameworks the learning curve for domain users is steep and the task of choosing and interacting with the right tools is highly difficult. Data analysis also requires that a data-analyst possess an integrated skill set spanning mathematics, computer-science, machine-learning, artificial-intelligence, statistics and a deep domain knowledge and understanding of the craft of problem formulation [27].

Another challenge that arises in the analysis of temporal data is in the inherent nature of the analytic process itself, which is typically a *computational, quantitative process*. A domain-user typically detects a pattern in the data (e.g., price jump) through computation of measures (e.g., stock return) and then applies these computed variables to several mathematical models and techniques. Analytic dilemma arises in the *planning of the analytic pipeline* when there is a choice of competing *variables* or *measures* that can be computed to detect a similar outcome and the choice of the measure dictates the computation tasks of the analysis pipeline. Translation of the complex computation and analytic model in the minds of the domain user into an analytic process is not a trivial task [27]. Data analysis can also be described as an *exploratory science* characterised by a cycle of model selection, customization and testing as scientists often do not know the exact goal or expected results beforehand. Data analysts cannot simply look at data and let the data speak for itself. They need to build models to interpret and gain the insight from the data. For example, an investment manager will be relying on a mathematical model to construct an optimal portfolio at a particular point in time. This model will use some underlying time-series variables that represent variations of asset risks and returns over time. The model can be back-tested by "populating variables" with data e.g. historical returns data. Depending on the performance and the accuracy of the model predictions, the user has several options: adjust the mathematical model, change some of the underlying variables or change the way data is mapped into the variables. The entire process is *iterative* in nature, characterised by repeated evaluations on new data-sets or by "tweaking" experimental parameters.

In light of the above challenges, the purpose of this paper is to propose a software architecture that facilitates analytics in a friendly, coherent and technology-agnostic manner. The rest of the paper is structured as follows. Section 2 provides the background and related work. Section 3 presents our solution which is based on a semantic reference model. In Sect. 4, we will apply the proposed model to enable analysts to identify price indicators of commodities. Section 5 concludes this paper.

2 Background and Related Work

Data analytics solutions have evolved from simple analytic techniques, along with supporting analytic tools, to sophisticated problem solving environments for data analysts. There are a wide range of analytic tools, techniques and problem-solving environments at the disposal of analysts. At one end of the spectrum software libraries provide programmers simple building blocks for building sophisticated analysis models and running experiments. There are a multitude of packages and libraries to leverage statistics, machine learning, text-mining, sentiment analysis, etc. [12]. A developer can build a program tailored to their needs utilizing libraries built using programming languages such as Java or Python or use "pre-packaged modules" such as those offered by the R programming language. It has been long argued that many end-users do not have, nor do they wish to, acquire programming skills just to use software packages effectively [13]. Data analysts have valuable skills and should spend their time doing science and not programming or data management. At the other end, application packages such as Microsoft Excel, Google Spreadsheets [20], Gnumeric [19], etc. offer simple, user-friendly interfaces which can be used to conduct elementary to intermediate level analyses. However, there are many shortcomings in such application packages; limited functions, rounding errors, miscalculations, etc. Overall, application packages are very handy tools for building models based on "clean data", but fail to cater to more sophisticated needs in data analysis.

In this paper, we take a higher level view, looking at high level design i.e. *an architecture* of solutions instead of specific solutions. A complete survey of existing approaches is presented in [11], where it has been noted that service-oriented and scientific workflow based environments are two key architectural approaches that have been extensively used by data scientists in coordinating processes for complex data analysis. *Service-oriented architectures* represent a strategy for the composition of distributed applications that propagates encapsulation of software artefacts as standards-based services. *Workflow based* approaches model work as a sequence of well-defined steps with a goal of providing a business service or performing a scientific experiment. In this approach each step corresponds to a single unit of work e.g., BPEL [15] and scientific workflow management systems such as Taverna [16], Kepler [17], Wings [18]. These existing approaches have focused on addressing essential criteria to support the activities of analysts, namely (1) automation and reproducibility (2) interoperability [27]. Most architectural strategies have aimed to provide support for a high level of *automation* in order to increase the efficiency and productivity of scientists by helping them to lower the effort and/or reduce time taken to complete tasks. The automation algorithms and code used by scientists contain concise information and instructions that are viewed as an accurate record of the research undertaken. The archived record of these instructions becomes a useful artefact for provenance tracking and ensures reproducibility of analysis processes. To enable the seamless interoperability of diverse tools and packages and enable the execution of process models on multiple platforms, *interoperability* has become another key focus. From this survey, it can be concluded that existing research efforts aimed at maximising automation, reproducibility and interoperability are quite mature, and have helped to mitigate analytic complexity. These approaches still fall short of addressing a third criterion, *usability*, the focus of which is to deliver an analytic approach to enable end-users to work in a more friendly and coherent manner in the face of extreme heterogeneity.

The closest related work to ours is ADAGE [9, 10, 14] which is an architectural framework that aims to achieve user-driven execution of processes through the composition of analysis functions as *services,* providing a reference event data model that enables ADAGE services to process data in a consistent manner. Design concepts underlying ADAGE aim to support data analysts by bringing together existing toolsets and data-sources, but are not adapted to capture the user's mental models and translate them into analytic pipelines. The ADAGE framework has been extended in [27], where a reference model was proposed to capture an event model and the user's analytic model with a specific domain. However, the reference model was not implemented using any standard formalisation. The main contribution of this paper is to explore the development of more complex semantic data models linked with existing ontologies that enable the standardisation of data formats as well as meaning and interpretation of data [13].

3 Proposed Approach

To support the design of analytic solutions tailored to the needs of non-technical data analysts we propose an architecture based on Model Driven Software Development and ADAGE principles, at the core of which lies a semantic reference model that represents an abstraction of the computational model to be applied on the data in order to gain insight. In this section, we first provide an overview of the design principles underlying our proposed architecture. Then, we describe semantic and statistical learning concepts. Finally, we present our ontology for modelling statistical learning algorithms.

3.1 Overview of Proposed Architecture

The proposed architecture can be seen as an application of the principles of *Domain Driven Design (DDD)* [21] and *Model Driven Software Development (MDSD)* [22] which emphasise that the heart of software development is knowledge of subject matter or *domain*. DDD is a design approach introduced by Eric Evans which focuses on creating a *model of the domain,* rather than the technology, using a high level of abstraction. This model should not just be a data schema or class diagram, but represents distilled knowledge about the domain and accurately expresses the behaviour of the domain through identifying important domain concepts and relationships between these concepts. The model shows how the domain users think of their domain problems in terms of these concepts. A domain model articulates domain problems and provides a practical approach to software design. It is equally important that the domain model is crafted carefully to enable it to be translated to practical implementations. A domain-driven design naturally leads to a model driven software development approach, which provides a software development approach to realise software systems from domain models. MDSD is currently a highly regarded software development paradigm, and a good fit for our proposed approach because of its *"consumer-centric"* or *"end-user-centric"* focus. MDSD approaches a software solution from the domain perspective, specifying that needed application functionality and behaviour be modelled formally in terms of the problem domain, without tainting it with technological concerns. MDSD focuses on *models* as central artefacts, where a model is a formal platform-independent module (PIM) that provides an abstract representation of a real-world application and applies *model transformations* to realise software systems from these PIMs. In our architecture, these models are defined as an ontology, captured using an appropriate semantic technology such as OWL [23], FIBO [24] etc.

Our proposed architecture defines an analytic stack that comprises of four main tiers, depicted in Fig. 2. At the lowest level of the stack is the *data* tier, which comprises different types of raw data to be analysed, such as structured and semi-structured data. The next tier is the *semantic reference model*, which is represented using an appropriate semantic technology such as OWL [23]. The semantic reference model encapsulates an *event model* and an *analytic model*. While the event model provides a standardisation of the representation of data from heterogeneous data-sources within a target domain, the analytic model captures the complex computation model in the minds of the domain user that is to be applied on the domain data. The tier above is the *analytics platform* layer, which encompasses the analytic tools that can be used to apply the analytic techniques embodied in the semantic reference model tier. The top tier is the higher-level *application layer* which can range from simple user interfaces and custom applications that produce visualisation results by running tools in the layer beneath to more sophisticated platforms such as scientific workflow management systems, which orchestrate a pipeline of analytic tasks to compose a single analytic process.

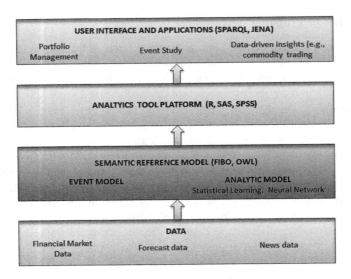

Fig. 2. Reference architecture

3.2 Semantic Reference Model: Basic Principles

The word *semantic* stands for 'meaning'; a *semantic concept* is a name used by the domain user to identify a domain object within a specific domain. For example stock price is a domain concept in the context of a financial application. *Ontology* describes a body of knowledge about a specific domain by defining the semantic concepts and semantic relationships between these concepts. Semantic relationships model the behaviour of the domain by capturing different kinds of associations between semantic concepts. To provide a formalisation to the vocabularies used in defining an ontology, W3C offers a large range of standard formats such as RDF and RDF schemas, Web Ontology Language (OWL) etc. For example, RDF represents information about the domain as *triples* which are a tuple of the form *<Subject, Predicate, Object>*, where *subject* and *object* represent two domain *semantic concepts* and *predicate* is a *semantic relationship* between these resources [28].

According to OASIS (Organization for the Advancement of Structured Information Standards) a *reference model* is:

> "an abstract framework for understanding significant relationships among the entities of some environment, and for the development of consistent standards or specifications supporting that environment. A reference model is based on a small number of unifying concepts and may be used as a basis for education and explaining standards to a non-specialist. A reference model is not directly tied to any standards, technologies or other concrete implementation details, but it does seek to provide a common semantics that can be used unambiguously across and between different implementations [2]."

The *semantic reference model* constitutes the core of our analytic stack. As previously noted, a vast amount of research and development has been directed towards analysing heterogeneous datasets, applying fragmented analytic techniques and building analytic models to unravel patterns. However, no generic model has been proposed to link the scattered knowledge in the data and the computational model in the minds of the domain-users. In proposing a semantic reference model we try to fill a major methodological gap, defining the semantics of a complex analytics model. As stated earlier, the semantic reference model comprises an *event model* and an *analytic model*. This paper focuses on defining an ontology to represent the complex analytic model and will rely on the event model proposed in [26]. Further, for the purposes of this paper, we limit the scope of our semantic analytic model to the analysis of data using statistical learning methods. The analytic model can be easily extended in the future to support other analytic techniques. Statistical learning addresses the general problem of function estimation based on empirical data, encompassing a wide array of popular algorithms and techniques for data analysis, pattern recognition and prediction [8].

Before we define our ontology for modelling statistical learning methods, we also define what statistical learning is using a simple example, as described by [25]. Suppose a researcher is interested to determine if the % change in inflation and the increase in population have an effect on beef consumption. In this context, the % inflation change and population increase are *independent variables* or *predictors* while beef consumption is the *dependent variable* or *response*. If the predictors are denoted as X_1, $X_2..X_n$ and the response is denoted as Y and Y is affected by the predictors, then we can define Y = f(X), where f is the function that connects the predictors X_1, $X_2..X_n$ to the response Y. This function, f, is generally unknown and one must estimate f based on observed data points. Statistical learning is a set of methods for estimating this function f. The two primary reasons for estimating f are *prediction* and *inference*. *Prediction* is about the using the estimated function f on a set of predictors, *X,* to calculate a predicted value for *Y*. *Inference* is concerned with how the response *Y* is affected as the predictors $\{X_1, X_2..Xn\}$ change. There are many linear and non-linear methods for estimating f and these methods can be broadly categorised as *parametric* and *non-parametric* methods [25]. We briefly provide an overview of parametric methods.

Given a set of data points or observations, these observations are referred to as *training data* as these observations will be used to train the method selected to estimate f. A parametric approach involves a two-step model based approach [25].

1. First, an assumption is made about the functional form or model of f, if f is linear in X it can be defined as:[1]

$$f(X) = \beta_0 + \beta_1 X_1 + \beta_2 X_2 + \beta_3 X_3 + \ldots + \beta_p X_p.$$

2. Once a model has been selected, the next step is to *fit* or *train* the model. In the previous step, if a linear model has been chosen the model estimator simply needs to estimate the parameters $\beta_0, \beta_1, \beta_2....\beta_p$, once values of these parameters have been estimated the function f is defined as,

---

[1] Function is multiple linear regression, which is a widely used form in statistical learning.

$$Y = f(X) = \beta_0 + \beta_1 X_1 + \beta_2 X_2 + \beta_3 X_3 + \ldots + \beta_p X_p.$$

One possible and quite commonly used approach to fitting the linear model is referred to as *ordinary least squares*. For our example, once the parameters have been estimated, we have a fitted linear model of the form:

beef consumption $= \beta_0 + \beta_1 * (\%\text{ inflation change}) + \beta_2 * (\%\text{population change})$

3.3 A Semantic Ontology for Modelling Statistical Learning

We now define our reference model which represents the key entities in statistical learning based on a parametric method. Tables 1 and 2 respectively define key semantic Classes and Properties used in the ontology.

This reference model is compatible with the Financial Industry Business Ontology (FIBO) [3]. The Financial Industry Business Ontology (FIBO) is a business conceptual ontology developed by the members of the Enterprise Data Management Council (EDM Council). FIBO provides a description of the structure and contractual obligations of financial instruments, legal entities and financial processes. FIBO is expressed

Table 1. Semantic classes for statistical learning ontology

Semantic ontology for statistical learning	
Semantic classes	Description
Functional form	The first step in a parametric based method is to assume a functional form or model for the function f e.g., a linear model or a non-linear model such as thin-plate spline
Measure	A variable e.g. beef consumption which could either be a predictor (independent variable) or a response (dependent variable)
Model estimator	A technique which given some measures, produces a **function** capable of predicting the values of one measure (dependent variable) based on the value of other measures (independent variables). So for our case study, the model estimator used is shown below e.g., Ordinary Least Squares estimator used to estimate parameters $\beta_0, \beta_1, \beta_2$
Fitted model	The fitted model obtained after the parameters have been estimated using an appropriate model estimator e.g., for our example, after applying the least squares estimator, *beef consumption* $= \beta_0 + \beta_1* (\%\text{ inflation change}) + \beta_2*(\%\text{population change})$ and if $\beta_0, \beta_1, \beta_2 = 2, 3, 4$ respectively then *beef consumption* $= 2 + 3* (\%\text{ inflation change}) + 4*(\%\text{population change})$ It can be seen for a given functional form assumed (e.g. Linear Model), different kinds of estimators (Least Squares, Lasso, etc.) can yield different fitted models
Function	Represents general functions such as $2 + 3x + 4y$ in the example above.

Table 2. Semantic properties for statistical learning ontology

Semantic properties	Description
generatesFittedModel	Definition: a predicate indicating any fitted model that is generated from model estimator by applying measures (training sets)
FMMeasureBinding (functional Model measure binding)	Definition: a predicate indicating the link between fitted model and the measure(s) in the fitted model.
MEMeasureBinding (model estimator measure binding)	Definition: a predicate indicating the link between model estimator and the measure(s) used it to estimate parameters.
determinesME	Definition: a predicate of Functional Form which indicates the form that is used for predicting a measure.
useFunction	Definition: a predicate that uses any function in mathematical definition.

in the RDF language of the Web (RDF/OWL) for machine readable interface processing and UML for human reading [3].

As shown in Fig. 3 our proposed *Reference Model* has five key classes: *Measure, FunctionalForm, Function, ModelEstimator and FittedModel.* These five classes together with ontologies that are part of FIBO capture key concepts of the reference model. The main concepts borrowed from FIBO include the entity *Measure* which represents an amount or degree of something; the dimensions, capacity or amount of something ascertained by measuring [5]. Measure is a subclass of the *Reference* ontology in FIBO. *Reference* is a concept that refers to (or stands in for) another concept. Every *Measure* is also a subclass of a FIBO *Thing* [4] which is a set theory construct.

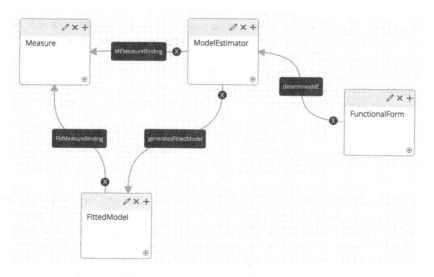

Fig. 3. Reference model in Jalapeno

The semantic reference model is implemented using the CAPSICUM framework, which provides meta-models, methods and tooling for developing dynamic, interactive business blueprints [6]. Users are able to maintain the reference model using Jalapeno, which is an interactive modelling platform for building CAPSICUM models. The models can be exported in a variety of formats (e.g. Turtle, RDF, XML, JSON, Marklogic Entity Model). FIBO ontologies have been imported into Jalapeno and are integrated in our reference model.

4 Commodity Pricing Case Study

This section describes a case study in which we explain the context (commodity pricing) and associated reference model, then describe a prototype implementation built following MDE principles.

4.1 Business Area and Associated Reference Model

The case study was inspired by a Hackathon organised at University of New South Wales in partnership with ANZ Bank in Australia. The motivation is that the future success of agribusiness will be reliant on informed decisions about capacity, investment and other driving factors. Many of the banks' customers are interested in questions like "which countries and consumers will buy our products? what prices and economic value is likely to be generated from this? what primary or processed food products should Australia seek to produce in future?". The idea of the competition was to use public and private data on this sector – macro-economic indicators, production volumes, weather patterns, prices, etc. to investigate what will drive this industry going forward [1].

Based on the available data, an instance of the analytics reference model was created to allow heterogeneous datasets to be analysed. Table 3 shows a sample of the measures used in the case study, the reference model would allow thousands of such measures to be defined (Table 4).

For example, applying a functional form "Multiple linear regression" to the measure *Beef and Veal export* (dependent variable) and the measures *Export of goods* and *Employment in agriculture* (independent variables) would produce a linear function of the form:

$$Beef\ and\ Veal\ export = F(Export\ of\ goods, Employment\ in\ agriculture)$$
$$= \beta_0 + \beta_1 Export\ of\ goods + \beta_2 Employment\ in\ agriculture$$

We have restricted ourselves to regression model estimators, so the models produced are regression functions (characterised by regression factors). Each of the models produced by a model estimator is an instance of a *FittedModel*. Given that there are potentially thousands of measures, it is possible to create millions of models each predicting a particular measure as a function of other measures. The measures themselves are connected (via FIBO ontologies) to other entities. For example, the Measure

Table 3. A sample of measures used in the case study

Measures
China - Beef and Veal export (KT)
China - Beef and Veal import (KT)
China - Beef and Veal Global Consumption
Beef price (us cents per pound)
China - Beef and veal global production (KT)
China - GDP per capita (Yuan)
China - Population (millions)
China - Import of goods (% change)
China - Export of goods (% change)
China - Inflation change (% change)
China - Gross national saving (% of GDP)
China - Per capita beef sold by rural household (kg)
China - Natural growth rate
China - Unemployment rate
Brent crude oil spot rate (USD per barrel)
China - Employment in agriculture (% of total employment)
China - Agricultural land (% of land area)
China - Urban population (% of total)

Table 4. Functional forms used in the case study

Functional form
Multiple linear regression
Multiple exponential regression
Multiple polynomial regression (second degree)
Stepwise regression

"China Beef and Veal Export" is linked to entities "Export" (FIBO Thing), "Beef and Veal" (FIBO EconomicResource) and "China" (FIBO Country). These are represented outside the scope of our case study but can be imported to enable more sophisticated usage of the reference model.

4.2 Prototype Implementation

Based on the general architecture presented in Sect. 3, we have built an analytics tool for identifying price indicators of commodities. The structure of the prototype is illustrated in Fig. 4. The tool has been developed in R and built using libraries such as Shiny and ShinyBS for the User Interface, MASS library to perform stepwise regression, Quandl to get data from quandl.com and XLConnect to read local dataset saved as excel files [28].

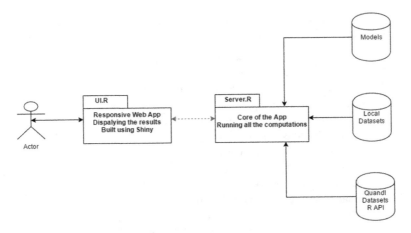

Fig. 4. Structure of the prototype

The tool works in two steps:

- A modelling step: where the user selects a variable to predict (dependent variable) and several possible explanatory variables (independent variables) in order to find the regression equation (model's equation) that describes the variations of the dependent variable.
- A forecasting step: using the equation of the first step and her own views on the independent variables, the user can forecast the dependent variable by inputting values for each of the independent variables which are fed into the model's equation to find a predicted value of the dependant variable.

In Fig. 5 we have provided a snapshot of the tool. The user interface is designed to enhance user interaction, we have grouped the measures by country and commodity. We have also provided an option for selecting models (or Model Estimator) to deploy an analytics model. The tool is equipped with a predictive section which uses the outcome of the analytics model to generate different scenarios. To analyse scenarios, the user can tweak the tolerance of the measures (Forecast Parameters) and select the type of forecasting model.

The structure of the user interface in the model leverages our Semantic Reference Model. All measures shown to users are the result of querying the reference model. In addition, R code is automatically generated from a user query. For example, the snippet below shows generic code in R that implements multiple linear regression once the user has selected the dependent and independent variables.

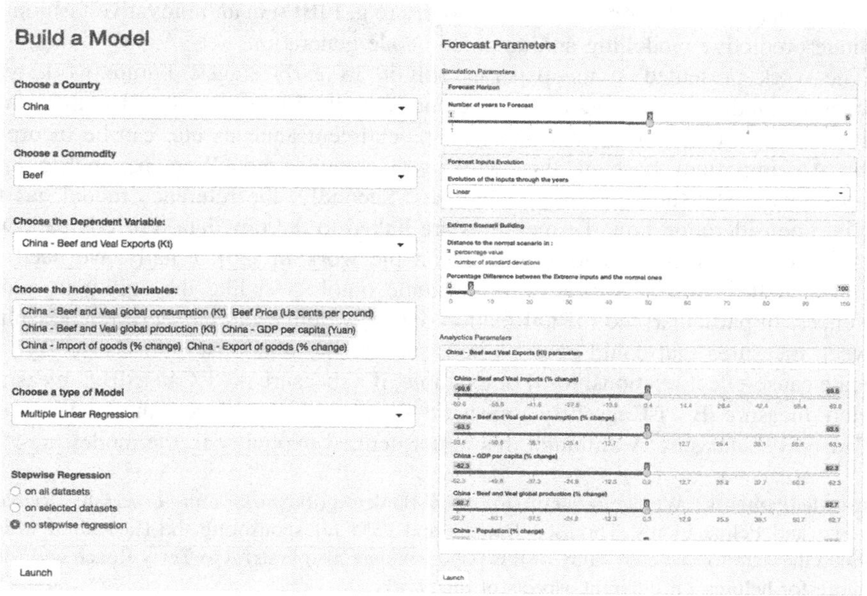

Fig. 5. Snap shots of commodity analytics tool

```
regressionModel <- function(data_set, dependent_var,
independent_var){

    fit <- lm(dependent_var~ independent_var)

return (fit)
}
```

Scalability is a property of this tool. Additional datasets can be added by modelling the appropriate measures in the reference model and such measures will be immediately available to the user via the User Interface. This architecture allows the user to create and add more analytics models or model estimators.

5 Conclusions and Future Work

This paper proposes a model-driven architecture that empowers domain experts to control and guide analytics processes in an exploratory way in the face of heterogeneity and complexity. The focus of this paper is a reference model which encompasses two main features: (1) a semantic model that captures the concepts in statistical learning algorithms and explicitly defines the relationships between variables, functional forms and model estimators, (2) leveraging statistical learning packages (e.g. R), semantic technologies (e.g. RDF) and existing ontologies (e.g. FIBO) in an innovative fashion to facilitate predictive modelling and automatic code generation.

The work presented in the paper is still in its early stages. Future work will concentrate on three areas. Firstly, the reference model will be generalized so that other analytics techniques such as text processing, sentiment analysis etc. can be incorporated. An important part of this work will include modelling parametric and non-parametric statistical learning techniques. Secondly, the reference model has to take into consideration how the measures are linked to the raw data. For this part, we intend to create an event ontology based on the work in [26]. Finally, we need to leverage all the constructs offered by semantic ontologies, like the ability to make inferences. In particular, the FIBO Relation ontology defines a rich set of relationships between measures that could be exploited, such as the relationship "isCausedBy" to indicate cause-effect relationships. For example, if a measure A "isCausedBy" measure B, and measure B "isCausedBy" measure C, it can be inferred that measure C "isCausedBy" measure A although this is not defined explicitly in the model.

Acknowledgements. We are grateful to ANZ Bank Agribusiness unit, especially Richard Schroder and Felipe Flores, Thomson Reuters and IBM for sponsoring the Hackathon which provided the data for the case study of this paper. We are also grateful to Terry Roach and Max Gillmore for helping on different aspects of this work.

References

1. Info Package for UNSW Data Science Hackathon. http://www.cse.unsw.edu.au/~fethir/HackathonInfo/HackathonStudentPack_v7.pdf. Accessed on 10 Sep 2016
2. OASIS SOA Reference Model Technical Committee. https://www.oasis-open.org/committees/tc_home.php?wg_abbrev=soa-rm/. Accessed on 10 Sep 2016
3. Financial Industry Business Ontology Foundations, The Enterprise Data Management Council. http://www.edmcouncil.org/edmcouncil. Accessed on 10 Sep 2016
4. Financial Industry Business Ontology (FIBO), Object Management Group. http://www.omg.org/spec/EDMC-FIBO/. Accessed on 10 Sep 2016
5. Merriam Webster, Measure (Definition). http://www.merriam-webster.com/dictionary/measure. Accessed on 10 Sep 2016
6. Roach, T.M.: CAPSICUM – A Semantic Framework for Strategically Aligned Business Architecture. Ph.D Thesis, UNSW, Sydney, Australia (2011)
7. Behnaz, A., Rabhi, F., Peat, M.: A software architecture for enabling time series analysis on real-time event data. In: Proceedings of International Work-Conference on Time Series, June 2016
8. Vapnik, V.: The Nature of Statistical Learning Theory, 2nd edn. Springer, New York (1999)
9. Rabhi, F.A., Yao, L., Guabtni, A.: ADAGE: a framework for supporting user-driven ad-hoc data analysis processes. Computing **94**(6), 489–519 (2012). doi:10.1007/s00607-012-0193-0
10. Yao, L., Rabhi, F.A.: Building architectures for data-intensive science using the adage framework. Concurrency Comput. Pract. Exp. **27**(5), 1188–1206 (2015)
11. Chen, J., Choudhary, A., Feldman, S., Hendrickson, B., Johnson, C., Mount, R., Sarkar, V., White, V., Williams, D.: Synergistic challenges in data-intensive science and exascale computing. DOE ASCAC Data Subcommittee Report, Department of Energy Office of Science (2013)

12. Yao, L., Rabhi, F., Peat, M.: Supporting data-intensive analysis processes: a review of enabling technologies and trends. In: Ramanathan, R., Raja, K. (eds.) Handbook of Research on Architectural Trends in Service-Driven Computing, vol. 2, pp. 481–508. IGI Global, Hershey (2014). doi:10.4018/978-1-4666-6178-3

13. Bernstien, P.A., Wecker, D., Krishnamurthy, A., Manocha, D., Gardner, J., Kolker, N., Reschke, C., Stombaugh, J., Vagata, P., Stewart, E.: Technology and data-intensive science in the beginning of the 21st century. Omics: J. Integr. Biol. **15**, 203–207 (2011)

14. Yao, L.: ADAGE A Framework For Supporting User-Driven Ad Hoc Data Analysis Processes. Doctor of Philosophy, University of New South Wales (2013)

15. OASIS, OASIS Web Services Business Process Execution Language (WSBPEL) TC | OASIS. https://www.oasis-open.org/committees/wsbpel/. Accessed 9 Sep 2016

16. TAVERNA 2009, Taverna - open source and domain independent Workflow Management System (2009). http://www.taverna.org.uk/. Accessed 9 Sep 2016

17. Tao, J., Zhao, Y.: Scientific workflow management and the Kepler system. Concurrency Comput. Pract. Exp. **18**, 1039–1065 (2006)

18. Deelman, E., Moody, J., Kim, J., Ratnakar, V., Gil, Y., Gonzalez-Calero, P.A., Groth, P.: Wings: intelligent workflow-based design of computational experiments. IEEE Intell. Syst. **26**(1), 62–72 (2011)

19. Gnumeric.org., Gnumeric (2016). http://www.gnumeric.org/. Accessed 17 Sep 2016

20. Apps.google.com. Google Sheets – Spreadsheets & Data Analysis for Business (2016). https://apps.google.com/intx/en_au/products/sheets/. Accessed 17 Sep 2016

21. Evans, E.: Domain-Driven Design: Tackling Complexity in the Heart of Software. Addison-Wesley Professional, Reading (2004)

22. Völter, M., Stahl, T., Bettin, J., Haase, A., Helsen, S.: Model-Driven Software Development: Technology, Engineering, Management. John Wiley & Sons, Hoboken (2013)

23. W3.org. OWL Web Ontology Language Guide. (2016). https://www.w3.org/TR/owl-guide/. Accessed 17 Sep 2016

24. W3.org. Financial Industry Business Ontology Community Group (2016). https://www.w3.org/community/fibo/. Accessed 17 Sep 2016

25. James, G., Witten, D., Hastie, T., Tibshirani, R.: An Introduction to Statistical Learning with Applications in R. Springer, New York (2013)

26. Milosevic, Z., Chen, W., Berry, A., Rabhi, F.A.: An open architecture for event-based analytics. Accepted in Int. J. Data Sci. Anal. (2016)

27. Natarajan, A.: Aventis, An architecture for event data analysis. Doctor of Philosophy, University of New South Wales (2016)

28. Behnaz, A., Rabhi, F., Peat, M.: A Software Architecture for Enabling Statistical Learning on Big Data. Springer Series on Statistics (2016)

Detecting Underwriters Stabilisation Trades: A Clinical Study

Qudamah Quboa[✉], Brahim Saadouni, Azar Shahgholian,
and Nikolay Mehandjiev

Alliance Manchester Business School, University of Manchester,
Manchester, UK
{Qudamah.Quboa,Brahim.Saadouni,Azar.Shahgholian,
N.Mehandjiev}@manchester.ac.uk

Abstract. In this study, we examine the stabilisation trades of United Rusal Company IPO's shares listed on the Hong Kong Stock Exchange (HKEx) and of its Global Depository Shares (GDS) that were simultaneously listed on Euronext Paris. Using both Thomson Reuters Tick History data and the HKEx rules and regulation relating to stabilisation, we identify and analyse the trades that were very likely to have been executed by the stabilisation manager (Credit Suisse) on both markets. We identify nearly 95% of the stabilisation trades on the Euronext Paris, with somewhat less accurate results for Hong Kong. Our results show that the stabilisation trades generated a profit equivalent to about 2.72% of the gross proceeds for the two lead underwriters, a profit which is bigger than their total underwriting commission of 2.31%.

Keywords: Ipos · Stabilisation · Underwriters · Process modelling · High frequency trades

1 Introduction

Price manipulation can be defined as artificially inflating or deflating the value of a security which is an illegal activity in all well-regulated and functioning stock markets, except the highly regulated price manipulation when underwriting banks stabilise prices during initial public offerings (IPOs). There is very limited research in this area despite the fact that price manipulation is viewed by regulators (both in developed and emerging markets) as very harmful since it adversely affects market integrity and distorts the allocation of capital. The misallocation of capital arises as the inflated or deflated stock prices no longer reflect the supply and demand for the affected securities. Stabilisation is highly regulated and hence permitted form of price manipulation as it allows the underwriter to limit the IPO market price falling well below the offer price.

Detecting trade-based manipulation during stabilisation period is very challenging for regulators. This is simply due to the uniqueness of executing buy or sells orders for securities without using asymmetric information.

B. Saadouni—This project was completed while Brahim Saadouni was holding a Visiting Fellowship at the School of Banking & Finance, University of New South Wales, Sydney, Australia.

S. Feuerriegel and D. Neumann (Eds.): FinanceCom 2016, LNBIP 276, pp. 32–46, 2017.
DOI: 10.1007/978-3-319-52764-2_3

Using High Frequency Trading (HFT) algorithms is a key current development in Financial Markets. This makes the detection of price manipulation much more complex [1]. HFT data is time stamped transaction-by-transaction or tick-by-tick data [2] which enable the discovery of prices and gains in spread and allow faster flow of information into prices. HFT data has unique characteristics which make the analysis of these data an interesting research challenge, especially the large volume of data to be processed, the existence of erroneous data and even disordered sequences and unspecified trade directions. These challenges impede the detection of the underwriters' trades during the stabilisation period of an IPO.

The study presented here focuses on identifying and analysing stabilisation trades using Thomson Reuters Tick History data and the Hong Kong rules and regulation concerning the stabilisation of the market prices of the shares of the IPO firms. The paper examines the stabilisation trades of one reputable underwriter (Credit Suisse acting as the stabilising manager) following the dual listing of United Rusal Company on the HKEx and Euronext Paris markets. The primary aim is to explore why the two main underwriters started with a short position of about 19% of the GDS and over-allocated about 13% of the Hong Kong shares despite the fact that the offering was not over-subscribed. We also aim to quantify the profits that the two lead underwriters generate from the stabilisation trades in the two markets.

The HKEx stock market disclosure requirements allow us to identify almost 95% of the underwriters (stabilisation) trades on Euronext Paris. However, our results for the Hong Kong market show that about 155% of the seller initiated trades could be classed as underwriters (stabilisation) trades. Further, we show that the underwriters profit from stabilisation amounts to about 2.72% of the gross proceeds raised by the IPO firm. This is over and above the 2.31% underwriting commission-gross spread- (including the incentive fee of 0.5%) earned by the two lead underwriters.

The rest of the paper is organised as follows. The next section briefly discusses the literature on IPO of securities and after market stabilisation. Section 3 introduces the methodology followed by detailed discussion of the relevant data of the case, then pre-processing and modelling steps are explained and finally, the discussion and the interpretation of the results are provided. Section 4 concludes the paper.

2   Literature Review

The consensus in the literature suggests that, on average, Initial Public Offerings (IPOs) are underpriced [3]. The IPO is one of the most important events in capital markets for any firm seeking listing. By providing access to public markets, the IPO is both the conduit for new capital to flow to fledgling companies and the mechanism for the existing owners to realize a return for their efforts [4]. One of the main IPO processes relates to the aftermarket trading activities of the lead underwriters after the official IPO listing. This is mainly concerned with price support (stabilisation) and overallotment option (OAO) which are the main focus of this study.

Regulators allow underwriters to stabilise the market prices of the shares of the IPO firms during the first thirty days of trading. The stabilisation enables the lead underwriters to delay or limit market prices to fall below the IPO offer prices. This facilitates the distribution of the shares of the IPO firms. In our paper, we focus on the main type of stabilisation. This involves the lead underwriter(s) often starting with a short position by over-allocating up to 15% more shares than what the issuing IPO firm wishes to sell. To protect the lead underwriters against market price being above the offer price, the over-allocation is always covered by the over-allotment option (OAO)[1] that the IPO firm grants to the underwriters prior to the first day of trading. The over-allotment option must be exercised within thirty days starting from the first day of trading. The OAO will only be exercised if the market price of the shares of the IPO firm is above the offer price. However, if the market price is equal to or below the offer price, the lead underwriter(s) can cover their short positions through market purchases. This form of stabilisation is the most prevalent in the vast majority of equity markets including for example, Hong Kong, Singapore, the UK and the US[2].

The Hong Kong Securities and Futures Commission (SFC) is responsible for the regulation relating to the market misconduct of the Hong Kong Stock Exchange. In Hong Kong, the regulation that prohibits stock market manipulation is the CAP 571 *Securities and Futures Ordinance* (Hong Kong) ('SFO'). This is defined under Sect. 245 of the SFO[3]. The Hong Kong Securities and Futures Commission (Price Stabilizing) Rules under the Securities and Futures Ordinance (SFO) permits the stabilizing manager (underwriter) to buy IPO shares in the secondary market in order to prevent or delay/limit possible decline of the market prices of the shares of the IPO firm during the first 30 days following official listing.

The empirical evidence relating to stabilisation provides several explanations for aftermarket price support. One paper [5] argues that IPO investors often base their investment decisions on their private information and the actions of other investors. Some investors may renege on their indications of interest if the market price of the IPO firm is below the offer price. Another source [6] argues that both stabilisation and underpricing can be used to prevent investors from cancelling their orders or reneging on their purchases by insuring that IPOs always start trading at or above their offer prices. There is also an argument [7] that uninformed investors face the "winner's curse"[4] as they are unable to distinguish between 'good' and 'bad' IPOs as informed investors do not bid if they believe the shares of the IPO firm are overpriced.

---

[1] In the US, the overallotment option is known as Green Shoe option after the first IPO firm to grant over-allocation powers to its underwriters (Green Shoe Manufacturing).

[2] In some other markets (e.g., the US) two other types of stabilisation might be used. These are: (1) Pure stabilisation: Underwriters are required to signal their intention to stabilise by posting a stabilising bid with a flag which can be identified by other market participants. Pure stabilisation is hardly used by US underwriters as it would reveal that the underwriters were unable to allocate all of the shares on offer; (2) the lead underwriters use what is known as penalty bids. This allows the lead underwriters not to pay selling concessions to the syndicate member(s) whose clients flip the shares of the IPO firm soon after listing.

[3] The rule covers insider dealing, false trading, price rigging, disclosure of false or misleading information and stock market manipulation.

[4] It is the retail (uninformed) investors who face this as the informed (institutional) investors will only apply for the shares of the IPO firms if the offer price is less than the intrinsic value of the stock. Thus, IPO shares must be offered at a discount to ensure that uninformed investors do not withdraw from the IPO market.

This results in the uninformed investors being allocated higher proportions of the shares of poor IPOs[5]. Therefore, underpricing[6] is used to compensate uninformed investors for adverse selection costs. The authors in [8] suggest that stabilization is more efficient than underpricing in compensating uninformed investors for the winner's curse. They argue that while both informed and uninformed investors benefit from underpricing, the benefits of stabilization tend to be targeted towards retail (uninformed) investors. Benveniste and Spindt [9] argue that rewarding investors who show strong indications of interest in the pre-offering stage with a higher allocation of underpriced stocks reduces information asymmetry among investors. However, receiving indications of interest also allows underwriters to gain an information advantage over investors. This new information asymmetry between underwriters and investors may provide underwriters with the incentive to overstate investor interest at the IPO. Benveniste et al. [10] argue that the information asymmetry problem that exists between underwriters and investors is mitigated if underwriters commit to supporting IPO prices in the aftermarket. Thus, their model indicates that underwriters stabilize IPO prices in order to protect their credibility with investors. In addition, in [11] the role of stabilization in protecting an underwriter's reputation with investors is emphasised. Overpricing an issue can damage an underwriter's reputation and decrease his future income (see [12]). Underwriters for this reason may choose to stabilize their offerings in order to mitigate investors' losses from purchasing overpriced IPOs. Finally, Fishe [13] argues that underwriters stabilize IPOs in order to maximize their own profits. Selling pressure from flippers[7] may cause the aftermarket price of an issue to drop below its offer price. Underwriters can respond to flippers by lowering the offer price, a strategy that reduces underwriting revenues. Alternatively, they may choose to begin trading with a short position and bring the issue to market, fully expecting its aftermarket price to decrease. This strategy allows an underwriter to make a profit by covering his short position at the lower aftermarket price. Fishe's model shows that underwriters respond to stock flippers by taking short naked positions during the pre-IPO period. Although these short positions allow underwriters to stabilize their offerings, the true purposes of stabilization, according to Fishe's model, are to maximize underwriters' profits and to penalize flippers rather than reduce investors' losses. Several studies show that underwriters stand ready to stabilise IPO prices by buying back shares in the aftermarket. The authors in [14] report that the level of underpricing associated with stabilised IPOs is lower than what it would have been in the absence of stabilisation. This finding suggests that stabilisation increases the benefits to the issuing firms in the form of lower underpricing. They also show that stabilisation is a significantly a valuable source of profitability for the underwriters. More specifically, they show that stabilisation enhances the overall profitability of the underwriters by an

---

[5] Poor IPOs are those firms that start trading at market prices below their offer prices.

[6] Underpricing is measured as the closing market price on the first day of listing minus the offer price divided by the offer price.

[7] This is the case of successful applicants who sell their allocated shares on the first few days of the IPO listing.

average of 29%. Overall, their results suggest that stabilisation is a better substitute for underpricing as it protects IPO investors from purchasing overpriced IPOs and brings benefits to both issuers and underwriters. In [11], stabilisation induces price rigidity at and below the offer price.

The paper examines the stabilisation trades of two lead underwriters in the Hong Kong and Euronext Paris markets. The Hong Kong regulation allows the stabilising manager to buy the IPO shares in the secondary market at or below the offer price in order to prevent or minimise a decline of the market price of the IPO firm. The regulation allows the underwriters to over-allocate (start with a short position) a maximum of 15% of the total issue size, exercise over-allotment option (OAO) to cover any short-position and liquidate any net long position in the IPO shares. The regulation also allows the underwriters to take a long position in the IPO firm providing that they can liquidate in an orderly fashion without distorting market prices. The rules allow a maximum period of 30 days for stabilization actions, which runs from the first day of the official listing and ends 30 days after either the closing date for subscription or the first day of trading whichever is the earlier.

Unlike the US where underwriters are not required to disclose whether they have stabilised the offering, the rules in Hong Kong compel the stabilising underwriters to disclose the following information within seven days from the expiry date of the stabilisation period: (1) if the offering was stabilised; (2) the expiry date of the sta-bilisation period; (3) the price range at which underwriters repurchased the shares (assuming there were more than one purchase for the purpose of stabilisation); (4) the extent to which the overallotment options were exercised; (5) the date of the last stabilisation trade and the price at which it was carried out. The Hong Kong regulation only allows stabilisation trades to be executed at or below the offer price. Furthermore, the stabilisation manager is not permitted to buy at a price above the previous stabil-isation price; (6) Announcement concerning the OAO. The OAO announcement should specify the date and the size if exercised; and (7) Underwriters and the issuing IPO Company are required to disclose full details as to whom (retail and institutional investors) the IPO shares are allocated and the extent to which the offer was over/undersubscribed (demand multiple). The IPOs company and its underwriters are also required to disclose the name(s) of major investors (e.g., sovereign wealth funds, cornerstone and strategic investors); amounts they have committed to invest; number of shares they will be allocated and their lock-up periods[8]. This latter disclosure is critical as it enables the lead underwriters to quantify the proportion of shares offered that cannot be sold within the first six months of trading. This was particularly important for the Rusal IPO case as the figures reported in Table 1 show that the top four investors acquired just over 39% of the shares offered. These investors had lock-up period of six-months.

---

[8] These three groups of investors tend to lock-up their investments for a minimum period of six months.

3 Proposed Approach

The proposed approach is following SEMMA data analysis method that is being developed by SAS Institute Inc. [15], including the processing steps of Sample, Explore, Modify, Model and Assess.

3.1 Sample and Explore: Data Understanding

United Company Rusal Limited IPO is selected as a case study. The company was listed on the HKEx and Euronext Paris stock market on the 27 January 2010. The lead underwriters were BNP Paribas & Credit Suisse where the latter was the main stabilising manager. The data relating to stabilisation and overallotment option announcements, the allocation/allotment of the shares and the company's prospectus are extracted from the HKEx website. The transaction and market depth data for the period 27th Jan 2010 through 19th Feb 2010 are obtained from Thomson Reuters Tick History database. Thomson Reuter's data is available in millisecond format. The initial sample includes 26,622 records of transactions for the IPO shares listed on the HKEx and 7,877 trades for the GDS listed on Euronext Paris.

The allocation data show that there are a total of 229 investors. 178 investors are allocated Hong Kong shares only, 48 investors are allocated GDS only and 3 investors are allocated both shares and GDS units. Details relating to the compensation of the underwriters were extracted from the company's prospectus. The data show that the underwriters are paid a fixed commission of about HK$315 million (this is equivalent to about 1.81% of the gross proceeds of HK$17.39 billion). The underwriters are also granted an incentive fee of 0.5% of the gross proceeds (this is worth about HK$87 million), the payment of which is at the discretion of the client company (in this case Rusal). Assuming the incentive fee is paid, the total underwriting commission is about 2.31%[9].

The Lead underwriters agreed to sell just over 1.6 billion of the company shares. The shares were sold in the form of ordinary shares in Hong Kong and Global Depository Shares in Euronext Paris. There are 20 shares per one GDS. The shares are priced at HK$10.80 each in Hong Kong and each GDS is priced at 19.91 Euros. Also, the underwriters are granted an OAO to sell an additional 225 million shares over and above the 1.6 billion shares on offer. The OAO gives the underwriters the option to sell an additional 225 million shares. In practice, the underwriters would over-sell (take a short position; in this case sell more than 1.6 billion shares, this is usually done when the demand is very strong) and then if (a) market price on the first 30 days after listing is above the offer price (HK$10.80 for the Hong Kong shares and Euros 19.91 in Euronext Paris) the underwriters would exercise the OAO (request the company to issue the additional 225 million shares); (b) if the market prices were below the offer price the underwriters would buy the 225 million shares (the underwriter is short) from

---

[9] The total commission of 2.31% (assuming the incentive fee is paid by United Company Rusal Ltd) that the lead underwriters received is the fixed commission of 1.81% plus the incentive fee of 0.5% of the gross proceeds.

the secondary market. The latter case enables the underwriters to stabilise the market prices by creating this short-term excess demand.

Table 1 Panel A reports the data relating to the allocation of the shares and the GDS to the top 25 investors. The figures show that the top 4 cornerstone[10] investors acquired about 39.44% of the offering. These investors had a lock-up period of six months. Panel A also reveals that the fifth investor is allocated just over 15% of the offering shares and the top 25 investors are allocated 87.42% of just over 1.6 billion shares of the IPO Company. Panel B shows the over-allocation figures for the Hong Kong shares and the GDS listed on Euronext Paris. The data show that the lead underwriters over-sell/over-allocate the entire 225 million shares included in the overallotment option. This is despite the fact that the demand for the shares is at best moderate. This over-allocation is about 13.62 of the Hong Kong shares, and just over 19% of the GDS in Euronext Paris. The data reveal that the two lead underwriters had a naked short position in Euronext Paris. The overall short position is about 14% of the total shares offered. This means that the risk of the naked position in the GDS (if the market price of the IPO firm is higher than the offer price) is non-existent as the underwriters can cover the short fall using the Hong Kong shares since their short position in the shares is only 14%. The profits from stabilisation trades will be discussed separately. Panel C of Table 1 shows the free float using the top 4, 5, 10, 25 investors respectively. The figures reveal that the percentage of shares that can be traded is about 1.34% as just over 87% of the shares on offer are allocated to 25 high net worth and cornerstone investors. These investors are unlikely to sell within the first few months of the listing of the IPO shares.

The determining of the data model is an important part of the conducted work to understand and to try to answer the research question. Figure 1 shows the overall data model of the United Company Rusal Ltd., the two lead underwriters (BNP Paribas & Credit Suisse), and the two stock markets (HKEx and Euronext Paris) in the high frequency data for this case study.

The data relating to the stabilisation announcement in the two markets are given in Table 2. The announcement provides details relating to the dates when the stabilising manager bought back the shares and the GDS of the IPO firm. The data also reveals the price ranges within which Credit Suisse acquired the shares during the period 27 January and 19 February in Hong Kong and 27 January to 11 February 2010 in Euronext Paris. One interesting observation is that the stabilisation period in Euronext Paris is shorter than the Hong Kong one. This is an interesting anomaly that is worth exploring as the price per share in the two markets should not be significantly different taking into account transaction costs and the exchange rate of the Hong Kong dollar vs Euro. We argue that this anomaly could be the result of the naked short position (lead underwriters over-selling) of 19% in the GDS. That is to say the lead underwriters created an excess supply of about 19% of the GDS. This over-supply was covered by aggressive purchases during the first few days following the listing of the GDS in

---

[10] Cornerstone investors include high net worth investors, sovereign wealth funds and corporate investors.

Table 1. Panel A provides details relating to the allocation of the number and percentage of shares allocated to the top 4, 5, 10 and 25 investors. Panel B provides details relating to the over-allocation in the two markets, HKEx and Euronext Paris separately. Panel C provides details relating to the free float based on top 4, 5, 10 and 25 investors, respectively. The free float is calculated as the difference between the total number of shares offered minus the total number of shares allocated to the top (4, 5, 10 or 25) investors divided by the total number of shares offered.

	No of shares/Global depository shares (figures in 000)	Allocated without over-allocation (%)
PANEL A		
Holdings of top 25 places		
Vnesheconombank (VEB)	477,090	29.63
NR Investment Ltd.	71,904	4.47
Paulson & Co. Inc.	71,736	4.45
Mr. Kuok Hock (Nien, Kerry Trading Co. Ltd., Cloud Nine Ltd. & Twin Turbo Ltd.)	14,376	0.89
Top 4 cornerstone investors	635,106	39.44
5th investor	243,720	15.14
Top 5 investor	878,826	54.58
Next 5 investors (6–10)	192,960	11.98
Top 10 investors	1,071,786	66.56
Next 15 investors (11–25)	336,000	20.87
Top 25 investors	1,407,786	87.42
Total offer (without over-allocation)	1,610,292.84	
Total offer (including over-allocation of 225)	1,835,293.84	
Issued & Fully paid capital	15,136,363.65	
PANEL B		
Cover in secondary market-HKEx & Paris	225,000	
Over-all short position (%)		13.97
Hong Kong		
Over-allocation (short-position)	204,815.62	
Total shares sold (excluding over-allocation)	1,504,158.38	
Total shares sold (including over-allocation)	1,708,974	
Hong Kong short position (%)		13.62
Euronext Paris		
Over-allocation (short-position in GDS)	1,009.24	
Total GDS sold (excluding over-allocation)	5,306.71	
Total GDS sold (including over-allocation)	6,315.94	

(*continued*)

Table 1. (*continued*)

	No of shares/Global depository shares (figures in 000)	Allocated without over-allocation (%)
Paris short position (%)		19.02
PANEL C		
Free float on day 1 using top 4 investors		6.44
Free float on day 1 using top 5 investors		4.83
Free float on day 1 using top 10 investors		3.56
Free float on day 1 top 25 investors		1.34

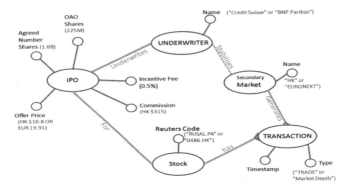

Fig. 1. The overall data model of the United Company Rusal Ltd., the two lead underwriters (BNP Paribas & Credit Suisse), and the two stock markets (HKEx and Euronext Paris) in the high frequency data.

Euronext Paris[11]. It is plausible that the underwriters may have intentionally over-sold the shares and the GDS to create the excess supply then they bought them back in the secondary market at prices below the two offer prices (HK$10.80 for the Hong Kong shares and Euros 19.91 for the GDS listed on the Euronext Paris) they sold them at.

To create a good financial dataset that could be used in the analysis, two steps are needed. Firstly, a reconstruction of full trades' records is required where each record will contain price, volume, ask price, and bid price in one complete record. Secondly, extra variables are required to be driven from the original variables which includes Mid-point price, various spread equations (quote, effective and realised spread), previous trade price, direction of trade (Seller or buyer initiated trades), and previous direction of trade.

---

[11] It is worth pointing out that one of the authors contacted Credit Suisse Hong Kong regarding the case and they declined to provide details relating to their stabilisation trades. We have also contacted the HKEx who also declined to provide any data relating to the case due to confidentiality clauses with its members.

Table 2. The table shows the dates and price ranges within which the stabilising managers traded in both markets (HKEx and Euronext Paris)

Hong Kong market (HK$)			Euronext Paris market (€)		
Date	Low price	High price	Date	Low price	High price
27-Jan-10	9.65	10	27-Jan-1	17.6	17.68
28-Jan-10	9.6	9.65	28-Jan-10	17.75	17.75
29-Jan-10	9.1	9.63	01-Feb-10	17.78	18
01-Feb-10	9.16	9.7	02-Feb-10	17.5	17.5
02-Feb-10	9.27	9.5	03-Feb-10	17.6	17.6
03-Feb-10	9.34	9.53	04-Feb-10	17.3	17.3
04-Feb-10	9.38	9.6	05-Feb-10	16.5	17
05-Feb-10	9.03	9.25	08-Feb-10	16	16.2
08-Feb-10	8.56	9.03	11-Feb-10	16.1	16.1
09-Feb-10	8.69	8.75			
10-Feb-10	8.71	8.78			
11-Feb-10	8.65	8.75			
12-Feb-10	8.44	8.66			
17-Feb-10	7.87	8.49			
18-Feb-10	7.37	8.01			
19-Feb-10	7.44	7.69			

3.2 Modify

After understanding the collected stock market data, the financial domain requirements, and what is missing and how it could be created, the data preparation step is initiated. This phase includes three main activities:

(1) Creating complete records of transaction trades from the raw data; this is done by completing the missing values (such as Bid price and Ask price) in the transactions where type variable is "Trade".
(2) Removing all extra records where the type variable is not "Trade".
(3) Cleaning extracted records from all transactions that does not have value in the "Volume" variable or contains "Trade cancellation" in the "Qualifiers" variable.
(4) Constructing and calculating new financial required variables: Mid-point price, quote spread, and previous trade price within the same day.
(5) Calculating direction of trade (Seller or buyer initiative), and previous direction of trade, within the same day using stock market trades classifications algorithms, which will be explained in details next.

Determining the trade direction is critical for our case as it allows us to distinguish between buyer and seller initiated trades. Our main focus is on the seller initiated trades since we are only interested in the transactions (both prices and volume) where the stabilising manager (Credit Suisse) is likely to be the buyer at the time of the transaction. The trade direction will also help in assessing the market's liquidity, the orders' imbalance, and the price of stabilisation. This classification of trades' direction is not provided by any of the high frequency trades' databases and therefore using trade classification algorithms is a necessity.

Most of the empirical literature on market microstructure use one or a combination of the following well-known algorithms: (a) the quote rule, (b) the tick rule, (c) the Lee-Ready (1991) rule. Firstly, the quote rule classification compares the trade price to the midpoint of the bid-ask price to classify any transaction. Secondly, the tick rule procedure is considering the price movements between the current and the previous transaction to calculate the transaction's initiative (buyer or seller). Lastly, the Lee-Ready algorithm is simply a combined algorithm of applying the quote rule algorithm first followed by the tick rule algorithm to have more accurate trade classification [16]. We use second-by-second trade and quotation data. The Thomson Reuters Tick History database does not provide any information whether the transaction is seller or buyer initiated. Therefore, we employ Lee and Ready (1991) algorithm to determine trade direction[12]. Any trade with a transaction price above the prevailing midpoint of quoted bid-ask spread is classified as a buyer initiated trade and vice versa. Any trade at the quoted midpoint is classified as seller initiated providing that the midpoint moved down from the previous transaction trade (downtick). In contrast, if the midpoint moved up (uptick), the transaction will be classified as a buyer initiated trade.

3.3   Model

To search and predict expected stabilisation trades activities, an understanding of the Hong Kong stabilisation rules and regulations is required. This starts with (1) looking on trades records that belongs to the security being analysed for each financial stock market separately; (2) considering only seller initiated trades *(trades direction = −1)*; (3) the dates of all trades must match with one of the announced stabilisation dates of the targeted stock market; (4) excluding all trades that does not belong to the announced price range of stabilisation dates; (5) marking the trades as stabilisation trades if *(the trades' price ≤ the previous valid trade price)*. The previous valid trade must belong to seller initiated trade that has a price > the minimum announced price range at specific stabilisation date. In addition, further analysis has been applied to provide enough support for domain experts to assess the behaviour of underwriters in financial stock market during the stabilisation period. This is done by (1) defining a new variable to describe the trade type (Seller normal trade, Seller stabilisation trade, Buyer trade); (2) calculating the time span between same type of trades; (3) driving financial statistics summary for each (specific and overall) trade type such as total number of trades, minimum and maximum volume and price, average price and volume, and total volume; and (4) counting the total number of shares (volume) of each type and comparing it with the underwriter's announced numbers. It should be noted that these summary statistics are calculated for each one of announced stabilisation dates of the targeted stock market.

---

[12] We also employ Wharton Research Data Services (WRDS) code for robustness checks.

3.4   Assess

The results reported in Table 3 show that on the first day of trading (27 January, 2010) about 88% of the seller initiated trades and 95% of the volume of trading on the Hong Kong market appear to have been undertaken by the stabilising manager. However, the figures for the GDS on the Euronext Paris were 78% and 73%, respectively. On day 12 of the stabilisation period (11 February, 2010), the stabilising manager appears to have been responsible for 90% of the number of seller imitated trades and 88% of the volume of trading on the HKEx. In contrast, the results for the Euronext Paris show that 25% of the number of trades and 8 of the volume are likely to have been executed by the stabilising manager. The results show that, on the first days of trading, the stabilising manager acquired the shares in Hong Kong at an average price of 9.15% below the offer price (average purchase price as a % of offer price of 90.85) and at 29.83% (70.17% of the offer price) below the offer price on the last day of stabilisation period (19 February, 2010). For the Euronext Paris, the stabilising manager acquired the GDS at 11.42% (88.58% of the offer price) below the offer price of 19.91 Euros per GDS on the first day of the stabilisation period and at 19.14% below the offer price (80.86% of the offer price) on the last day (11 February, 2010)of the stabilisation period.

The results relating to the imbalance between the buyer and seller initiated trades and the stabilisation profit in the two markets are reported in Table 4. We also report the imbalance between the seller and buyer initiated trades for the Hong Kong and the Euronext Paris markets and the average stabilisation price in the Hong Kong and the Euronext Paris in Fig. 2. The results for the Hong Kong market show that the stabilising manager makes an average profit of about HK$445 Million[13]. Figure 2 reveals that seller initiated trades exceeded buyer initiated trades during most of the stabilisation period (27 January to 19 February, 2010). Figure 2 also shows the average stabilisation price that stabilisation manager pays for the Hong Kong shares during the stabilisation period. The results show that the price ranges from a high of HK$10.00 on the first to a low of 7.44 on the last day of the stabilisation period. The price decline mirrors the imbalance between seller and buyer initiated trades. The results for Euronext Paris show that the stabilising manager made an average of profit 2.55 million Euros (this is equivalent to about HK$27.5 million). In addition, Fig. 2 reports the imbalance between seller and buyer initiated trades on the Euronext Paris and the average price that stabilising manager pays for the GDS on Euronext Paris. The figures show that trades undertaken during the first few trading days (up to the 3 February, 2010) on Euronext Paris stabilised the GDS prices at about 17.6 Euros vs. an offer price of 19.91 Euros per GDS and the GDS price fell to a low of 16.1. Overall, our results show that the stabilisation manager may have made a total profit from stabilisation trades of about HK$472.5 million. This is equivalent to about 2.72% of the gross proceeds of about HK$17.39 billion that the company raised from the offering.

---

[13] Our results for the Hong Kong market may have over-estimated the profits that Credit Suisse made in the Hong Kong market as were not able to identify the stabilisation trades with a high degree of certainty. Our figures over-estimate the number of shares that Credit Suisse bought back by about 33%. Thus, the profit may have been HK$298.15 million.

Table 3. The table provides details relating the actual price within which the stabilising manager traded, total number of seller initiated trades and the total volume of seller initiated trades, details relating to the stabilisation and non-stabilisation trades (e.g. total number of trades (%), total and average volume (%), average price as % of the offer and minimum and maximum volume number of traded shares per day. Same data is provided for non-stabilisation trades. The data is given for both the Hong Kong market and Euronext Paris.

HKEx Market

Date	Price Low	Price High	Seller Initiated Total No of Trades	Seller Initiated Total Volume (100)	Stab. Total No of Trades %	Stab. Total Vol %	Stab. Avg. Vol. (100)	Stab. Avg. Price% of offer price	Stab. Min. Vol. (100)	Stab. Max. Vol. (100)	Non-Stab. Total No. of Trades%	Non-Stab. Total Vol.%	Non-Stab. Avg. Vol. (100)	Non-Stab. Avg. Price% of offer price	Non-Stab. Min. Vol. (100)	Non-Stab. Max. Vol. (100)
27-Jan-10	9.65	10.00	874.00	1,351,480.00	88	95	1,656.71	90.85	40	185,760	12	5	0.04	91.19	40	6,480
28-Jan-10	9.60	9.65	416.00	721,894.72	92	62	1,160.84	89.06	40	49,920	8	38	0.43	89.44	80	160,000
29-Jan-10	9.10	9.63	255.00	387,780.00	89	88	1,511.59	87.66	100	24,960	11	12	0.08	87.19	100	9,600
01-Feb-10	9.16	9.70	301.00	190,792.00	85	89	658.33	86.98	232	6,240	15	11	0.03	87.25	240	2,400
02-Feb-10	9.27	9.50	176.00	261,120.00	81	83	1,508.81	86.69	240	10,560	19	17	0.07	86.68	240	9,120
03-Feb-10	9.34	9.53	140.00	101,840.00	80	60	545.00	87.36	80	2,640	20	40	0.07	87.09	80	10,800
04-Feb-10	9.38	9.60	119.00	66,640.00	89	85	533.58	88.30	120	3,840	11	15	0.04	88.34	240	2,400
05-Feb-10	9.03	9.25	177.00	97,240.00	91	92	553.54	84.38	40	5,520	9	8	0.03	84.54	200	4,080
08-Feb-10	8.56	9.03	117.00	72,420.00	89	91	636.35	81.19	60	5,280	11	9	0.02	80.94	240	1,440
09-Feb-10	8.69	8.75	102.00	48,840.00	86	90	496.82	80.69	40	3,600	14	10	0.02	80.68	80	960
10-Feb-10	8.71	8.78	54.00	41,818.00	78	75	748.57	81.05	240	6,000	22	25	0.04	80.62	40	3,120
11-Feb-10	8.65	8.75	73.00	57,040.00	90	88	763.64	80.48	240	3,600	10	12	0.05	80.55	160	2,640
12-Feb-10	8.44	8.66	91.00	35,040.00	96	97	391.72	79.45	240	1,680	4	3	0.01	78.78	240	240
17-Feb-10	7.87	8.49	249.00	124,540.00	85	87	515.17	75.17	20	6,240	15	13	0.02	75.22	240	2,400
18-Feb-10	7.37	8.01	340.00	149,370.00	86	85	437.42	70.41	20	5,520	14	15	0.02	70.16	240	3,120
19-Feb-10	7.44	7.69	173.00	77,040.00	69	73	474.62	70.17	160	3,360	31	27	0.02	70.55	160	3,600

Euronext Paris Market

Date	Price Low	Price High	Seller Initiated Total No of Trades	Seller Initiated Total Volume (100)	Stab. Total No of Trades %	Stab. Total Vol %	Stab. Avg. Vol. (100)	Stab. Avg. Price% of offer price	Stab. Min. Vol. (100)	Stab. Max. Vol. (100)	Non-Stab. Total No. of Trades%	Non-Stab. Total Vol.%	Non-Stab. Avg. Vol. (100)	Non-Stab. Avg. Price% of offer price	Non-Stab. Min. Vol. (100)	Non-Stab. Max. Vol. (100)
27-Jan-10	17.60	17.68	199.00	8,294.48	78	73	39.05	88.58	0.01	600	22	27	50.93	88.54	0.01	400
28-Jan-10	17.75	17.75	34.00	1,589.07	59	88	70.01	89.15	0.10	400	41	12	13.50	89.63	0.03	50
01-Feb-10	17.78	18.00	47.00	1,389.27	68	77	33.46	89.94	3.10	100	32	23	21.23	90.20	4.00	50
02-Feb-10	17.50	17.50	82.00	1,479.89	13	16	21.64	87.90	5	50	87	84	17.49	88.82	0.50	92
03-Feb-10	17.60	17.60	22.00	489.80	23	63	62.00	88.40	10	100	77	37	10.58	88.52	0.80	40
04-Feb-10	17.30	17.30	18.00	346.92	72	89	23.85	86.89	3.70	69	28	11	7.36	86.95	2.00	20
05-Feb-10	16.50	17.00	45.00	785.32	93	95	17.79	83.61	0.52	136	7	5	12.67	83.12	3.00	30
08-Feb-10	16.00	16.20	42.00	666.33	50	73	23.22	80.68	0.75	176	50	27	8.51	82.11	0.75	30
11-Feb-10	16.10	16.10	12.00	358.24	25	8	9.46	80.86	6.05	11	75	92	36.65	81.07	1.16	140

Table 4. Table provides details relating to the imbalance between buyer and seller initiated trades both in Hong Kong and Paris, profits from stabilisation in the two markets, total stabilisation volume in the two markets and offer prices.

HKEx market		Euronext Paris market	
Total number of shares		Total number of shares	
Buyer initiated trades	133,103,740.00	Buyer initiated trades	1,077,933.00
Seller initiated trades	378,489,472.00	Seller initiated trades	1,539,932.00
Trade imbalance	− 245,385,732.00	Trade imbalance	− 461,999.00
Total profit	445,137,991.20	Total profit	2,545,573.46
Total stabilisation volume:	317,682,200.00	Total stabilisation volume:	1,064,585.00
Offer price	HK$10.80	Offer price	Euros 19.91

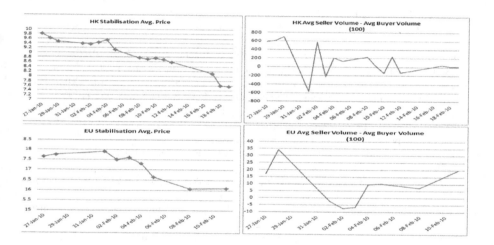

Fig. 2. The trading average prices and the imbalance in the trading volumes during the stabilisation period for seller initiated trades in both the HKEx and the EU markets.

The results show that the underwriters can profit from stabilisation to the extent the profit is greater than the underwriting commission.

4 Conclusion

We examine the stabilisation of the IPO shares of United Company Rusal Ltd. that was listed on both the HKEx and Euronext Paris. The lead underwriters were PNB Paribas and Credit Suisse. Credit Suisse was appointed as the stabilising manager. The Hong regulation requires the underwriters to disclose details relating to the over-allotment options and also to whether the offering was stabilised. Using high frequency trading data that is extracted from Thomson Reuters Tick Database, we are able to identify the trades that are likely to be the stabilisation trades. The findings show that we are able to identify about 95% of the stabilisation trades of the GDS on the Euronext Paris. However, our finding for the HKEx market is less accurate. We find that just over 317 million seller initiated trades can be classed as stabilisation trades vs. an over-allocation of over 204 million Hong Kong shares. Further, the over-allocation in the two markets enables the two lead underwriters to buy back the shares at a profit which is equivalent to 2.72% of the gross proceeds raised by the IPO firm. This is higher than the 2.31% underwriting commission that the lead underwriters earned from the offering.

References

1. Antonios, A.: Stock Market and Economic Growth: An Empirical Analysis for Germany. Bus Econ. J., 1–12 (2010)
2. Engle, R.F.: The econometrics of ultra-high-frequency data. Econometrica **68**(1), 1–22 (2000)
3. Loughran, T., Ritter, J.R.: The new issues puzzle. J. Financ. **50**(1), 23–51 (1995)
4. Ellis, K., Michaely, R.: O'hara, M.: When the underwriter is the market maker: An examination of trading in the IPO aftermarket. J. Financ. **55**(3), 1039–1074 (2000)
5. Welch, I.: Sequential sales, learning and cascades. J. Financ. **47**(2), 695–732 (1992)
6. Schultz, P.H., Zaman, M.A.: Aftermarket support and underpricing of initial public offerings. J. Financ. Econ. **35**(2), 199–219 (1994)
7. Rock, K.: Why new issues are underpriced. J. Financ. Econ. **15,** 187–212 (1986)
8. Chowdhry, B., Nanda, V.: Stabilization, syndication, and pricing of IPOs. J. Financ. Quant. Anal. **31**(1), 25–42 (1996)
9. Benveniste, L.M., Spindt, P.A.: How investment bankers determine the offer price and allocation of new issues? J. Financ. Econ. **24**(2), 343–361 (1989)
10. Benveniste, L.M., Busaba, W.Y., Wilhelm, W.J.: Price stabilization as a bonding mechanism in new equity issues. J. Financ. Econ. **42**(2), 223–255 (1996)
11. Lewellen, K.: Risk, reputation, and IPO price support. J. Financ. **61**(2), 613–653 (2006)
12. Nanda, V., Yun, Y.: Reputation and financial intermediation: An empirical investigation of the impact of IPO mispricing on underwriter market value. J. Financ. Intermed. **6**(1), 39–63 (1997)
13. Fishe, R.P.H.: How stock flippers affect IPO pricing and stabilization. J. Financ. Quant. Anal. **37**(2), 319–340 (2002)
14. Mazouz, K., Agyei-Ampomah, S., Saadouni, B., Yin, S.: Stabilization and the aftermarket prices of initial public offerings. Rev. Quant. Financ. Acc. **41**, 417–439 (2013)
15. Dean, J.: Big Data, Data Mining, and Machine Learning: Value Creation for Business Leaders and Practitioners. Wiley, New York (2014)
16. Ellis, K., Michaely, R., O'Hara, M.: The accuracy of trade classification rules: Evidence from NASDAQ. J. Financ. Quant. Anal. **35**(04), 529–551 (2000)

Predicting Corporate Credit Ratings Using Content Analysis of Annual Reports – A Naïve Bayesian Network Approach

Petr Hajek[(✉)], Vladimir Olej, and Ondrej Prochazka

Institute of System Engineering and Informatics, Faculty of Economics
and Administration, University of Pardubice,
Studentska 84, 532 10 Pardubice, Czech Republic
{petr.hajek,vladimir.olej,ondrej.prochazka}@upce.cz

Abstract. Corporate credit ratings are based on a variety of information, including financial statements, annual reports, management interviews, etc. Financial indicators are critical to evaluate corporate creditworthiness. However, little is known about how qualitative information hidden in firm-related documents manifests in credit rating process. To address this issue, this study aims to develop a methodology for extracting topical content from firm-related documents using latent semantic analysis. This information is integrated with traditional financial indicators into a multi-class corporate credit rating prediction model. Informative indicators are obtained using a correlation-based filter in the process of feature selection. We demonstrate that Naïve Bayesian networks perform statistically equivalent to other machine learning methods in terms of classification performance. We further show that the "red flag" values obtained using Naïve Bayesian networks may indicate a low credit quality (non-investment rating classes) of firms. These findings can be particularly important for investors, banks and market regulators.

Keywords: Credit rating · Firms · Prediction · Concept extraction · Naïve Bayesian network

1 Introduction

Corporate credit ratings are intended to provide capital market participants with an evaluation for comparing the creditworthiness (capability and willingness of a firm to meet its payable commitments). The evaluation is particularly important for investors (institutional and individual), banks and market regulators, because it measures a default risk in a benchmark fashion. According to rating agencies such as Moody's, Standard & Poor's or Fitch, a credit rating is reported to require a variety of information necessary for the final evaluation. This information includes financial statements, corporate annual & quarterly reports, conference calls, management interviews, etc. The information is processed by a group of experts to reach an objective and independent rating grade (usually on a rating scale from Aaa/AAA denoting the highest credit quality to D representing default or bankruptcy).

© Springer International Publishing AG 2017
S. Feuerriegel and D. Neumann (Eds.): FinanceCom 2016, LNBIP 276, pp. 47–61, 2017.
DOI: 10.1007/978-3-319-52764-2_4

In corporate default prediction literature, previous work has mainly focused on two approaches, structural and empirical [1]. The structural approach aims to model default probability based on the underlying dynamics of interest rates and firm-related indicators such as market capitalization [2]. In the empirical approach, on the other hand, the model is learned from data. The research has tended to focus either on the estimation of default probability or two-class bankruptcy prediction. This is mainly due to the specific characteristics of rating predictions such as the ordinal scaling of rating grades and multi-class prediction. Imbalanced classes are another issue to be addressed. As a result, it is difficult to measure the performance of prediction models.

The corporate credit rating is a time-consuming and expensive process, requiring an in-depth expert analysis of the underlying information. In recent years, there has been therefore an increasing interest in simulating the credit rating process of rating agencies through machine learning methods (e.g. [3–16]). These methods include hidden Markov models [3], neural networks [4, 5], support vector machines [6–9], decision trees [10], fuzzy systems [11], rough sets [12], hybrid systems [13, 14], and meta-learning approaches [15, 16].

However, a major problem with this kind of application is the selection and accessibility of input variables (credit rating determinants). The main limitation of the above-mentioned studies is the focus on financial ratios (such as profitability, liability, or liquidity), which can be easily obtained from corporate financial statements.

Given the results of the studies (see e.g. [8] for a summary), it appears that the level of information available in financial data is bounded, resulting into a maximum of 80% accuracy for a multi-class problem [6]. This suggests that additional input variables are required to obtain significantly better results. This is also in line with the methodologies of rating agencies that emphasize the importance of qualitative factors in their credit rating process. Additionally, information extracted from firm-related textual documents have shown promising prediction ability recently. Specifically, the relative frequency of selected word categories in annual reports such as positive/negative sentiment [17] and modality/certainty/activity [18] have shown highly predictive abilities. Similarly, negative sentiment in news articles was reported more important for future credit rating changes compared with positive sentiment [19, 20]. The research to date has tended to examine predefined word categories (dictionaries related to overall sentiment/opinion) rather than the topical content of textual documents. For related bankruptcy prediction problem, Cecchini et al. [21] extracted words with highest relative frequency from corporate annual reports and performed the detection of synonymous words using WordNet ontology. However, the above-mentioned studies failed to detect the structures and links between the concepts in firm-related textual documents. To bridge this gap, this study was aimed to develop a methodology for extracting topical content from firm-related documents. We developed this methodology to examine the importance of firm-related textual concepts in the highly imbalanced ordered multi-class problem of rating prediction.

This information is combined with traditional financial indicators to predict corporate credit rating. We demonstrate that although financial indicators are critical to predict rating grades, textual information may increase prediction performance. We believe that this approach may contribute to a greater understanding of the linguistic character of firm-related textual documents. In addition, the application of Naïve

Bayesian networks enables developing the "red flag" values of predictive variables indicating the presence of a low credit quality. In contrast to other machine learning methods, Naïve Bayesian networks can be considered as probabilistic white-box classifiers, facilitating the understanding of complex relationships within the data through probability distributions [22]. As far as we know, such probability distributions have not been reported in the literature. Subsequently, they can also be used to better model default probability in the structural models.

The remainder of this paper has been divided into four sections. The paper first gives an overview of the research methodology applied to predict corporate credit ratings. Specifically, Sect. 2 lays out the theoretical foundations of textual content analysis and Naïve Bayesian networks, respectively. Section 3 describes the result of the content analysis of corporate annual reports. Section 4 provides the results of experiments and analyzes the performance of the proposed approach. Finally, Sect. 5 concludes this paper and discusses its implications.

2   Research Methodology

The research methodology (depicted in Fig. 1) includes collection and pre-processing of both financial indicators and text information. The relevancy of pre-processed words in a particular document were obtained using a traditional *tf.idf* (term frequency weighted with inverse document frequency) weighting scheme, where the relative frequency of a word in a document is compared to the inverse proportion of the word over the entire corpus of documents [23]. The application of latent semantic analysis led to a lower-dimensional semantic space, where topic analysis of the corpus could be performed. Then, the two categories of variables (textual and financial) were integrated into one prediction model, which consisted of feature selection and classification into rating classes.

2.1   Financial Indicators

Rating agencies do not make the determinants of corporate credit rating public. However, their methodologies suggest that financial indicators represent important factors in the corporate credit rating process. In previous studies (see e.g. [10] for a review), broader categories such as profitability, liquidity, leverage, and market value ratios are usually considered as the most important financial ratios.

For this study, a set of 35 financial indicators was drawn from the Value Line Database and Standard & Poor's database for 557 U.S. firms (mining and financial companies were excluded from the dataset since they require specific financial indicators). As presented in Table 1, the set included: (1) size of firms; (2) corporate reputation; (3) industry membership; (4) profitability ratios; (5) activity ratios; (6) business situation; (7) asset structure; (8) liquidity ratios; (9) leverage ratios; and (10) market value ratios. Data for all the financial indicators were collected for the year 2010.

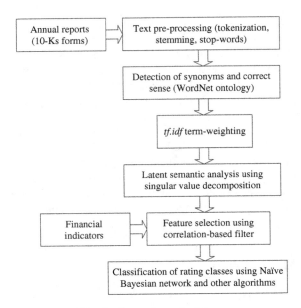

Fig. 1. Reseach methodology.

Table 1. List of financial indicators used in this study.

Category	Indicators
Size of firms	Total assets, sales, cash flow, enterprise value
Corporate reputation	Shares held by mutual funds, shares held by insiders
Industry membership	Standard Industrial Classification code
Profitability ratios	Return on assets, return on equity, return on capital, operating margin, net margin, enterprise value/earnings before interest, taxes, depreciation and amortization
Activity ratios	Sales/total assets, operating revenue/total assets
Business situation	Effective tax rate, sales growth
Asset structure	Share of fixed assets within total assets, share of intangible assets within total assets, non-cash working capital, working capital/total assets
Liquidity ratios	Current ratio, cash ratio
Leverage ratios	Book debt/total capital, market capitalization/total debts, market debt/total capital, net gearing
Market value ratios	Dividend yield, 3-year stock price variation, beta, earnings per share, stock price/earnings, payout ratio, price-to-book value, high/low stock price

The firms were labelled with rating classes obtained from the Standard & Poor's rating agency in the year 2011. The rating classes are defined on the rating scale AAA, AA, ..., D. Figure 2 depicts the rating classes along with their frequencies in the dataset. Rating classes BBB, BB and B prevailed in the dataset, whereas rating classes C and D were not present at all. The frequencies also suggest a highly imbalanced classification problem.

Following recent studies on corporate credit rating prediction, we used feature selection procedure to include only informative financial indicators. Feature selection was also shown to improve the prediction performance of classification models in prior literature [10]. In order to provide the same subset of financial indicators for all classification algorithms, we used a correlation-based filter that optimizes the set of input variables by considering the individual predictive ability of each variables along with the degree of redundancy between the variables [24]. The correlation-based filter was chosen mainly because of the ordinal scaling of rating grades. Specifically, the rating grades were treated as the problem of ordinal classification. To avoid overfitting and feature selection bias, we used 10-fold cross-validation and performed the feature selection procedure only on training data, this is 10 times. All financial variables selected at least once are presented in Table 2 together with their mean values.

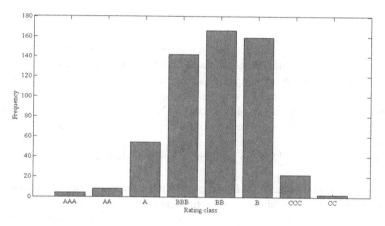

Fig. 2. Frequencies of rating classes in dataset.

2.2   Latent Semantic Analysis for Concept Extraction

Documents are usually represented in a bag-of-words fashion (only the frequency of words matters, their order is ignored) with a very high dimensionality (each word representing one variable). However, a lower-dimensional semantic space is favorable for topic analysis. This dimensionality reduction can be performed using two general approaches, latent semantic analysis (using singular value decomposition - SVD) and probabilistic topic models (such as probabilistic latent semantic analysis or latent dirichlet allocation). We used latent semantic analysis in order to obtain an interpretable semantic model. In latent semantic analysis, semantic space is constructed

Table 2. Mean values of selected financial indicators for rating classes.

Indicator	AAA	AA	A	BBB	BB	B	CCC	CC
Revenues	120.8	52.2	27.9	12.3	4.2	3.4	1.5	0.6
ROE	0.30	0.10	0.64	0.22	0.24	−0.03	−0.42	−0.51
MD/TC	0.03	0.14	0.16	0.26	0.34	0.53	0.60	1.00
EPS	3.64	4.73	3.25	3.09	2.04	0.85	−0.33	NA
High/Low	0.21	0.30	0.28	0.33	0.43	0.54	0.65	0.80
3yr stock var.	0.21	0.25	0.30	0.35	0.50	0.70	0.99	1.41
PR	0.41	0.53	0.52	0.99	0.40	0.31	0.04	NA
Dividend yield	0.03	0.04	0.03	0.06	0.02	0.02	0.02	0.00

Legend: ROE – return on equity, MD/TC – market debt to total capital, EPS – earnings per share, High/Low – high/low stock price, 3yr stock var. – 3-year stock price variation, PR – payout ratio, NA – missing value.

from the SVD of the term-document matrix. In this new space, documents with the same concepts but different terms can be found [25].

SVD [26] is the factorization of the term-document matrix $\mathbf{X}$, which have m lines (terms) and n columns (documents), into

$$\mathbf{X} = \mathbf{U} \Sigma \mathbf{V}^T, \tag{1}$$

where $\mathbf{U}$ (m x m dimension) and $\mathbf{V}$ is (n x n dimension) are orthonormal matrixes and Σ (m x n dimension) is diagonal (diagonal values are the singular values of the matrix $\mathbf{X}$). The columns of $\mathbf{U}$ are the left singular vectors of the matrix $\mathbf{X}$, and the columns of $\mathbf{V}$ (or the rows of $\mathbf{V}^T$) are the right singular vectors. To compute the SVD is to find the eigenvalues and the eigenvectors of $\mathbf{X}\mathbf{X}^T$ and $\mathbf{X}^T\mathbf{X}$, where the eigenvectors of $\mathbf{X}^T\mathbf{X}$ are the columns of $\mathbf{V}$ and the eigenvectors of $\mathbf{X}\mathbf{X}^T$ are the columns of $\mathbf{U}$. The singular values of $\mathbf{X}$ in the diagonal of matrix Σ are the square root of the common positive eigenvalues of $\mathbf{X}\mathbf{X}^T$ and $\mathbf{X}^T\mathbf{X}$. The number of positive singular values equals the rank of the matrix.

In the term-document matrix $\mathbf{X}$, it is important to select an appropriate term frequency weighting scheme because simply using the term frequency tends to exaggerate the contribution of the terms [25]. Commonly used term-weighting schemes, such as *tf.idf*, can address this issue.

2.3 Naïve Bayesian Networks

Naïve Bayesian Networks (also known as Bayesian Networks and Bayesian Belief Networks) are probabilistic graphical models that represent knowledge about an uncertain domain [27, 28]. Naïve Bayesian Networks consist of a set of nodes and set of directed edges between the nodes. Both the nodes and directed edges form a directed acyclic graph G. The nodes represent random variables. The edges represent direct dependences between the variables. The variables have a finite set of mutually exclusive states. All interdependencies are described using conditional probability distributions. To each variable with parents there is attached a probability table. Naïve Bayesian Networks are based on Bayes' theorem so that they can reason against the causal direction. Formally, a Naïve Bayesian Network B defines a unique joint probability distribution P over a set of random variables $\mathbf{U}$ [29] as follows:

$$P_B(X_1, \ldots, X_n) = \prod_{i=1}^{n} P_B(X_i | \Pi_{X_i}) = \prod_{i=1}^{n} \theta_{X_i | \Pi_{X_i}}, \tag{2}$$

where $X_1, \ldots, X_n$ are random variables, and Θ represents the set of parameters that quantifies the network. Thus, independence assumption is encoded in graph G, this is each variable X_i is independent of its non-descendants given its parents in G.

Naïve Bayesian Networks are used to reason under uncertainty. They can estimate certainties for the values of variables that are not observed or their observation is very costly. They are also used as a representation for encoding uncertain expert knowledge in expert systems [30]. This is usually done by learning Naïve Bayesian Networks from data in order to induce a network that best fits the probability distribution over the set of training data. Heuristic search algorithms such as hill climbing, genetic algorithm or simulated annealing are used to find the optimum structure.

3   Content Analysis of Corporate Annual Reports

The annual reports (10-Ks forms) of corresponding 557 U.S. firms were collected at the U.S. Securities and Exchange Commission EDGAR System. The corpus of 557 filings was the result of document collection and pre-processing. The average document size (in number of characters) was 496,183. We downloaded all annual reports in txt format (without amended documents) for the year 2010. Documents that only referred to other reports were withdrawn. Similarly, graphics, tables and SEC header were removed from the documents before text pre-processing. First, all words were converted to lower case letters. Further, linguistic pre-processing included tokenization, stemming (Snowball stemmer) and discarding the stop-words in the corpus (using the Rainbow stop-word handler). Next, the potential term candidates were compared with the WordNet ontology [31] to detect synonyms and the correct sense of the terms for the domain (those with the highest score for Economy, Commerce or Law domains were chosen following [32]).

To represent the weights (term-weighting scheme) of the pre-processed words (i.e. how important a word is within a document), we used $tf.idf$ as the most common approach. In this scheme, weights w_{ij} are calculated as follows:

$$w_{ij} = \begin{cases} (1 + \log(tf_{ij})) \log \frac{N}{df_i} & \text{if } tf_{ij} \geq 1 \\ 0 & \text{otherwise} \end{cases}, \tag{3}$$

where N denotes the total number of documents, tf_{ij} is the frequency of the i-th word in the j-th document, and df_i denotes the number of documents with at least one occurrence of the i-th term. To select the most relevant words, we ranked them according to their *tf.idf* and used the top 1000 for our experiments. The most relevant 1000 words are usually enough to discriminate document categories from each other [33, 34].

To extract the topical concepts from the corpus of corporate annual reports, we performed SVD and chose those concepts with singular values greater than 1 (76 concepts with the maximum singular value of 48.13), see Fig. 3. Further, the concepts had to be labeled based on the term importance. In the resulting vector space, semantic concepts can be interpreted due to the semantic relatedness between terms (they are placed near one another). Each term can be characterized by a weight indicating the strength of the semantic association. In other words, the concepts represent extracted common meaning components. Table 3 presents the concepts with the highest singular values along with the most important terms (largest weights). The meanings (labels) were manually assigned to the concepts based on the semantic association.

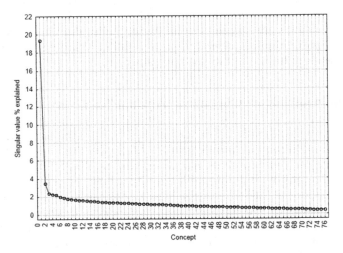

Fig. 3. Singular value explained by extracted topical concepts.

Similarly as for the financial indicators, only informative concepts were used in subsequent corporate credit rating prediction. Therefore, the correlation-based filter was used to optimize the set of concepts. The following concepts were selected at least once: (1) corporate restructuring; (2) investment policy; (3) financial restructuring; and (4) domestic market difficulties. The mean values of the concepts for each rating class are presented in Table 4. These value suggest that firms with low credit quality mention corporate and financial restructuring less frequently in their annual reports. On the other hand, they used words related to investment policy and domestic market difficulties more frequently.

Table 3. Labels of topical concepts and representative words.

Label of concept	Most important words
Corporate restructuring	Restructur, manufactur, swap, currenc, redempt, …
Relation to environment	Manufactur, inventori, environment, labor, long-liv, …
Investment policy	Indentur, indebted, construct, lender, libor, …
Financial restructuring	Remedi, bond, court, alleg, lawsuit, …
Legal proceedings	Court, alleg, licens, lawsuit, violat, …
Legal proc. implications	Court, alleg, lawsuit, restructur, redempt, …
Debt policy	Redempt, indentur, stock-bas, real, bond, …
Financial coop. and partnership	Partner, currenc, merger, enterpris, third-parti, …
Foreign markets	Polit, foreign, convert, countri, emerg, …
Domestic market difficulties	American, unfavor, cancel, forc, downturn, …
E-commerce	Internet, space, billion, center, expans, …
…	…

Table 4. Mean values of selected topical concepts for rating classes.

Concept	AAA	AA	A	BBB	BB	B	CCC	CC
Corporate restructuring	0.159	0.021	0.018	0.022	0.022	0.022	0.022	0.023
Investment policy	−0.024	−0.019	−0.014	−0.003	0.006	0.011	0.016	0.005
Financial restructuring	0.007	0.017	0.014	0.014	−0.004	−0.009	−0.011	−0.040
Domestic market difficulties	−0.021	−0.008	−0.011	−0.006	−0.001	0.004	0.008	−0.004

4 Experimental Results

To predict corporate credit rating using the combination of financial indicators and extracted concepts, we employed Naïve Bayesian network. To compare its performance, we performed the experiments also for several commonly used machine learning methods such as decision trees (C4.5 and Random Forest), neural networks (multilayer perceptron - MLP) and support vector machines (sequential minimum optimization algorithm - SMO), as well as statistical methods (logistic regression and k-nearest neighbor classifier). As stated above, we used 10-fold cross-validation to avoid over-fitting.

The methods were trained using the settings presented in Table 5. Naïve Bayesian network was trained using several heuristic search algorithms, namely a hill climbing algorithm, K2 (a hill climbing algorithm restricted by an order on the variables), genetic algorithm, and simulated annealing. Bayes scoring function was used to measure the quality of a network structure.

Common classification performance criteria such as accuracy may lead to misleading conclusions for imbalanced datasets [35]. Measures such as ROC (receiver operating characteristic) curve have been reported more appropriate for imbalanced datasets. We adopted this approach and measured the quality of prediction using the area under the ROC curve. A ROC is a graphical plot which illustrates the performance of a binary classifier system. Therefore, it is necessary to calculate average ROC across all classes to measure the overall performance of the methods. As reported by [36], the ROC measure has no obvious generalisation to multiple classes. However, it can be approximated by averaging the set of two-dimensional ROCs. Here we used a 1-vs-rest approach where ROC is weighted by class probability estimates [37].

Table 5. Settings of machine learning methods.

Method	Parameters and their values
Naïve Bayesian network	Hill climbing algorithm (no. of parents = {1, 2})
	K2 (no. of parents = {1, 2})
	Genetic algorithm (descendant population size = 10, population of network structures = 5, and no. of generations = 10)
	Simulated annealing (start temperature = 10, decreasing factor delta = 0.999, and no. of iterations = 10000)
C4.5	Minimum no. of instances per leaf = 2, and confidence factor for pruning = 0.25
Random Forest	Maximum depth of trees unlimited, no. of trees to be generated = {100, 200, 400}, and no. of variables randomly sampled as candidates at each split = $\log_2(\#predictors) + 1$
SMO	Complexity parameter $C = \{2^0, 2^1, 2^2, ..., 2^5\}$, polynomial kernel function with exponent = {1, 2}, RBF kernel function with gamma = 0.01
MLP	Neurons in hidden layer = $\{2^2, 2^3, 2^4\}$, learning rate = 0.1, and no. of iterations = 500
Logistic regression	Broyden–Fletcher–Goldfarb–Shanno learning algorithm
k-nearest neighbor	k = {3, 5} neighbors

Table 6 shows that Naïve Bayesian network performed statistically similar to Random Forest, SMO and MLP. The best network structure was found by the K2 hill climbing algorithm with 1 parent (represented by the output variable - rating class). In order to assess the impact of financial and textual indicators, additional experiments were conducted without considering these sets of variables. Table 7 presents the results obtained from the sensitivity analysis. Results from this table can be compared with the results in Table 6. As can be seen from Table 7, the performance of most classifiers significantly decreased when the financial indicators were not used. This result confirms their crucial importance in predicting corporate credit ratings. In contrast, no significant differences were found between the performances on all indicators and those on financial indicators (without textual indicators). However, the performance of the best classifiers (Naïve Bayesian network and Random Forest) decreased without using textual indicators, suggesting a limited information hidden in the text of corporate annual reports.

Table 6. Classification performance in terms of average ROC.

Method	Mean ± St.Dev.	t-value (p-value)
Naïve Bayesian network	0.9237 ± 0.0751	
C4.5	0.6051 ± 0.2428	6.889 (0.000)[a]
Random Forest	0.9252 ± 0.1747	−0.419 (0.676)
SMO	0.8702 ± 0.2240	1.663 (0.100)
MLP	0.8754 ± 0.2086	1.642 (0.105)
Logistic regression	0.8174 ± 0.3005	2.280 (0.025)[a]
k-nearest neighbor	0.6204 ± 0.2183	8.653 (0.000)[a]

Legend: [a]significantly worse than Naïve Bayesian network with all indicators at $p = 0.05$ using Student's paired t-test.

Table 7. Classification performance in terms of ROC (Mean ± St.Dev.) for datasets without financial and textual indicators.

Method	Without financial	Without textual
Naïve Bayesian network	0.7284 ± 0.2208[a]	0.9221 ± 0.0653
C4.5	0.5674 ± 0.2458[a]	0.6365 ± 0.2634[a]
Random Forest	0.6543 ± 0.2832[a]	0.9188 ± 0.1895
SMO	0.8621 ± 0.0912	0.8691 ± 0.2266
MLP	0.7608 ± 0.3614[a]	0.9107 ± 0.0912
Logistic regression	0.9066 ± 0.2090	0.9207 ± 0.0679
k-nearest neighbor	0.4929 ± 0.0100[a]	0.7455 ± 0.2528[a]

Legend: [a]significantly worse than Naïve Bayesian network with all indicators at $p = 0.05$ using Student's paired t-test.

In Table 8, the probability distributions are presented for the input variables (note that only average values are shown across 10-fold cross-validation). The probabilities were merged to two values (> value/ ≤ value) where more than two nodes were present for a variable. For example, ROE ≤ 0.128 is a "red flag" value, indicating the presence of a low credit quality ($P = 1-0.82 = 0.18$ for AAA class, ..., $P = 0.77$ for CCC class and $P = 0.57$ for CC class). Low rating classes (in this case BB, ..., CC) are also known as non-investment rating classes. In fact, Table 8 indicates a strong change in probability distributions for this category of rating classes. Notably, the probability of non-investment rating class sharply increases for Revenues ≤ 8675, EPS ≤ 1.47, High/Low > 0.416, PR ≤ 0.168, dividend yield ≤ 0.008, financial restructuring ≤ 0.022 and domestic market difficulties > −0.010. Smaller changes can be observed inside the investment (AAA, ..., BBB) and non-investment rating classes, respectively.

Table 8. Probability distribution for rating classes.

Variable	Value	AAA	AA	A	BBB	BB	B	CCC	CC
Revenues	>8675	0.82	0.79	0.62	0.34	0.08	0.05	0.02	0.14
ROE	>0.128	0.82	0.79	0.93	0.62	0.54	0.32	0.23	0.43
MD/TC	>0.312	0.18	0.21	0.11	0.36	0.47	0.83	0.77	0.57
EPS	>1.47	0.82	0.89	0.89	0.88	0.62	0.38	0.28	0.71
High/Low	>0.416	0.18	0.21	0.04	0.18	0.51	0.76	0.81	0.57
3yr stock var.	>0.327	0.25	0.35	0.40	0.56	0.86	1.00	0.98	0.88
PR	>0.168	0.90	0.72	0.75	0.73	0.31	0.13	0.07	0.17
Dividend yield	>0.008	0.90	0.72	0.88	0.80	0.43	0.32	0.33	0.50
Corporate restructuring	–	–	–	–	–	–	–	–	–
Investment policy	> −0.009	0.30	0.17	0.43	0.62	0.82	0.89	0.93	0.83
Financial restructuring	>0.022	0.30	0.39	0.30	0.37	0.10	0.03	0.07	0.17
Domestic market difficulties	> −0.010	0.10	0.39	0.59	0.58	0.72	0.86	0.76	0.83

Legend: – variable not selected as informative in Naïve Bayesian networks.

5   Conclusion

This paper has given an account of and the reasons for the widespread use of textual analysis in corporate credit rating prediction. Specific topics such as investment and financial policy seem to be particularly important for credit rating assignment. The evidence from this study also suggests that capital market participants should pay attention to unusually low/high values of selected informative indicators.

The purpose of the current study was to design a methodology for extracting topical content from corporate annual reports. The methodology can also be applied to other firm-related documents such as earnings press releases, conference calls, news stories, analyst disclosures, and social media. Potential applications of topical content analysis include the prediction of future earnings, stock returns, volatility, financial fraud, etc. However, this study was limited to traditional latent semantic analysis mainly because we aimed to extract an easy-to-interpret semantic space. Probabilistic topic models, on the other hand, can be further extended to investigate other linguistic structures. Further investigation and experimentation into alternative topic models is therefore strongly recommended. For example, latent dirichlet allocation has recently been applied to extract topics from corporate press releases [38]. In addition, future studies should deal with the strong imbalance of credit rating datasets. Finally, to further our research we are planning to integrate the topic model with sentiment analysis to extract more informative indicators from firm-related documents.

The experiments in this study were carried out in Statistica 12 and Weka 3.7.13 using the MS Windows 7 operation system.

Acknowledgments. This work was supported by the scientific research project of the Czech Sciences Foundation Grant No: GA16-19590S and by the grant No. SGS_2016_023 of the Student Grant Competition.

References

1. Atiya, A.F.: Bankruptcy prediction for credit risk using neural networks: a survey and new results. IEEE Trans. Neural Networks **12**(4), 929–935 (2001). doi:10.1109/72.935101
2. Crouhy, M., Galai, D., Mark, R.: A comparative analysis of current credit risk models. J. Bank. Finance **24**(1–2), 59–117 (2000). doi:10.1016/S0378-4266(99)00053-9
3. Petropoulos, A., Chatzis, S.P., Xanthopoulos, S.: A novel corporate credit rating system based on Student's-t hidden Markov models. Expert Syst. Appl. **53**, 87–105 (2016). doi:10.1016/j.eswa.2016.01.015
4. Zhong, H., Miao, C., Shen, Z., Feng, Y.: Comparing the learning effectiveness of BP, ELM, I-ELM, and SVM for corporate credit ratings. Neurocomputing **128**, 285–295 (2014). doi:10.1016/j.neucom.2013.02.054
5. Hajek, P.: Municipal credit rating modelling by neural networks. Decis. Support Syst. **51**(1), 108–118 (2011). doi:10.1016/j.dss.2010.11.033
6. Huang, Z., Chen, H., Hsu, C.J., Chen, W.H., Wu, S.: Credit rating analysis with support vector machines and neural networks: a market comparative study. Decis. Support Syst. **37**(4), 543–558 (2004). doi:10.1016/S0167-9236(03)00086-1
7. Kim, K.J., Ahn, H.: A corporate credit rating model using multi-class support vector machines with an ordinal pairwise partitioning approach. Comput. Oper. Res. **39**(8), 1800–1811 (2012). doi:10.1016/j.cor.2011.06.023
8. Hajek, P., Olej, V.: Credit rating modelling by kernel-based approaches with supervised and semi-supervised learning. Neural Comput. Appl. **20**(6), 761–773 (2011). doi:10.1007/s00521-010-0495-0
9. Chen, C.C., Li, S.T.: Credit rating with a monotonicity-constrained support vector machine model. Expert Syst. Appl. **41**(16), 7235–7247 (2014). doi:10.1016/j.eswa.2014.05.035
10. Hajek, P., Michalak, K.: Feature selection in corporate credit rating prediction. Knowl.-Based Syst. **51**, 72–84 (2013). doi:10.1016/j.knosys.2013.07.008
11. Hajek, P.: Credit rating analysis using adaptive fuzzy rule-based systems: an industry-specific approach. Cent. Eur. J. Oper. Res. **20**(3), 421–434 (2012). doi:10.1007/s10100-011-0229-0
12. Chen, Y.S., Cheng, C.H.: Hybrid models based on rough set classifiers for setting credit rating decision rules in the global banking industry. Knowl.-Based Syst. **39**, 224–239 (2013). doi:10.1016/j.knosys.2012.11.004
13. Wu, T.C., Hsu, M.F.: Credit risk assessment and decision making by a fusion approach. Knowl.-Based Syst. **35**, 102–110 (2012). doi:10.1016/j.knosys.2012.04.025
14. Yeh, C.C., Lin, F., Hsu, C.Y.: A hybrid KMV model, random forests and rough set theory approach for credit rating. Knowl.-Based Syst. **33**, 166–172 (2012). doi:10.1016/j.knosys.2012.04.004
15. Pai, P.F., Tan, Y.S., Hsu, M.F.: Credit rating analysis by the decision-tree support vector machine with ensemble strategies. Int. J. Fuzzy Syst. **17**(4), 521–530 (2015). doi:10.1007/s40815-015-0063-y
16. Hájek, P., Olej, V.: Predicting firms' credit ratings using ensembles of artificial immune systems and machine learning – an over-sampling approach. In: Iliadis, L., Maglogiannis, I., Papadopoulos, H. (eds.) AIAI 2014. IAICT, vol. 436, pp. 29–38. Springer, Heidelberg (2014). doi:10.1007/978-3-662-44654-6_3
17. Hájek, P., Olej, V.: Evaluating sentiment in annual reports for financial distress prediction using neural networks and support vector machines. In: Iliadis, L., Papadopoulos, H., Jayne, C. (eds.) EANN 2013. CCIS, vol. 384, pp. 1–10. Springer, Heidelberg (2013). doi:10.1007/978-3-642-41016-1_1

18. Hajek, P., Olej, V., Myskova, R.: Forecasting corporate financial performance using sentiment in annual reports for stakeholders' decision-making. Technol. Econ. Dev. Econ. **20**(4), 721–738 (2014). doi:10.3846/20294913.2014.979456

19. Lu, Y.C., Shen, C.H., Wei, Y.C.: Revisiting early warning signals of corporate credit default using linguistic analysis. Pacifin-Basin Finan. J. **24**, 1–21 (2013). doi:10.1016/j.pacfin.2013. 02.002

20. Lu, H.M., Tsai, F.T., Chen, H., Hung, M.W., Li, S.H.: Credit rating change modeling using news and financial ratios. ACM Trans. Manag. Inf. Syst. **3**(3), 14 (2012). doi:10.1145/ 2361256.2361259

21. Cecchini, M., Aytug, H., Koehler, G.J., Pathak, P.: Making words work: using financial text as a predictor of financial events. Decis. Support Syst. **50**(1), 164–175 (2010). doi:10.1016/j. dss.2010.07.012

22. Dejaeger, K., Verbraken, T., Baesens, B.: Toward comprehensible software fault prediction models using Bayesian network classifiers. IEEE Trans. Software Eng. **39**(2), 237–257 (2013). doi:10.1109/TSE.2012.20

23. Salton, G., Buckley, C.: Term-weighting approaches in automatic text retrieval. Inf. Process. Manag. **24**(5), 513–523 (1988). doi:10.1016/0306-4573(88)90021-0

24. Yu, L., Liu, H.: Feature selection for high-dimensional data: a fast correlation-based filter solution. In: International Conference on Machine Learning, ICML 2003, Washington, vol. 3, pp. 856–863 (2003)

25. Crain, S.P., Zhou, K., Yang, S.H., Zha, H.: Dimensionality reduction and topic modeling: from latent semantic indexing to latent dirichlet allocation and beyond. In: Aggarwal, C.C., Zhai, C. (eds.) Mining Text Data, pp. 129–161. Springer, New York (2012). doi:10.1007/ 978-1-4614-3223-4_5

26. Wall, M.E., Rechtsteiner, A., Rocha, L.M.: Singular value decomposition and principal component analysis. In: Berrar, D.P., Dubitzky, W., Granzow, M. (eds) A Practical Approach to Microarray Data Analysis, pp. 91–109. Kluwer (2003). doi:10.1007/0-306- 47815-3_5

27. Howard, R.A., Matheson, J.E.: Influence diagrams. Decis. Anal. **2**(3), 721–762 (2005). doi:10.1287/deca.1050.0020

28. Pearl, J.: Probabilistic Reasoning in Intelligent Systems: Networks of Plausible Inference. Morgan Kaufmann, San Mateo (1988)

29. Heckerman, D., Geiger, D., Chickering, D.M.: Learning Bayesian networks: the combination of knowledge and statistical data. Mach. Learn. **20**(3), 197–243 (1995). doi:10.1007/ BF00994016

30. Friedman, N., Geiger, D., Goldszmidt, M.: Bayesian network classifiers. Mach. Learn. **29**(2–3), 131–163 (1997). doi:10.1023/A:1007465528199

31. Miller, G.A.: WordNet: a lexical database for English. Commun. ACM **38**(11), 39–41 (1995)

32. Hajek, P., Olej, V.: Comparing corporate financial performance and qualitative information from annual reports using self-organizing maps. In: 10th International Conference on Natural Computation (ICNC 2014), pp. 93–98. IEEE (2014). doi:10.1109/ICNC.2014.6975816

33. Matveeva, I., Levow, G.-A., Farahat, A., Royer, C.H.: Term representation with generalized latent semantic analysis. In: Recent Advances in Natural Language Processing IV: Selected papers from RANLP 2005, Current Issues in Linguistic Theory, vol. 292, pp. 45–54. John Benjamins Publishing (2007)

34. Hájek, P., Boháčová, J.: Predicting abnormal bank stock returns using textual analysis of annual reports – a neural network approach. In: Jayne, C., Iliadis, L. (eds.) EANN 2016. CCIS, vol. 629, pp. 67–78. Springer, Heidelberg (2016). doi:10.1007/978-3-319-44188-7_5

35. Chawla, N.V., Japkowicz, N., Kotcz, A.: Editorial: special issue on learning from imbalanced data sets. ACM Sigkdd Explor. Newsl. **6**(1), 1–6 (2004)
36. Hand, D.J., Till, R.J.: A simple generalisation of the area under the ROC curve for multiple class classification problems. Mach. Learn. **45**, 171–186 (2001). doi:10.1023/A:1010920819831
37. Provost, F., Fawcett, T.: Robust classification for imprecise environments. Mach. Learn. **42**(3), 203–231 (2001). doi:10.1023/A:1007601015854
38. Feuerriegel, S., Ratku, A., Neumann, D.: Analysis of how underlying topics in financial news affect stock prices using latent dirichlet allocation. In: Bui, T.X., Sprague, R.H. (eds) 49th Hawaii International Conference on System Sciences (HICSS), pp. 1072–1081. IEEE (2016). doi:10.1109/HICSS.2016.137

Say It at the Right Time: Publication Time of Financial News

Dorina Palade, Simon Alfano$^{(\boxtimes)}$, and Dirk Neumann

University of Freiburg, Freiburg, Germany
`simon.alfano@is.uni-freiburg.de`

Abstract. By law, stock-listed companies must immediately disclose any information that might influence the valuation of the company in order to ensure a fair supply of information to all interested parties. However, laws and regulations do not specify clear requirements regarding the language used and the exact timing when information may be considered relevant enough for disclosure. Previous research shows that delaying bad news provides more time to adjust the language of an announcement in order to encourage a more optimistic perception. This paper investigates how the positive or negative character of news content influences the daily timing of the announcement and how the timing relates to stock performance. We find that negative messages are slightly longer than positive ones. In addition, announcements released before trading tend to have a more positive sentiment than those released during intraday trading, which may reflect a longer preparation time.

Keywords: Behavioral finance · News sentiment · Publication timing · Sentiment analysis

1 Introduction

Financial markets are an integral part of the economy and a driving factor in boosting economic development. As economic research has revealed, financial markets are not only highly competitive and volatile, but also driven by behavioral and partly non-rational factors [26]. For instance, Barberis emphasizes that behavioral anomalies and strategic behavior, such as how to frame and when to signal information, increasingly influence finance research [19]. He argues that regulators should update their rules, so that they account for anomalies in the decision-making process.

In this context, we conduct a case study on the German stock market. In Germany, stock-listed companies must immediately disclose any information that might influence the valuation of their company per German regulations as stated in the Wertpapierhandelsgesetz (WpHG). This legal measure prevents insider trading and ensures a fair supply of information to all the interested parties [1]. However, the laws and regulations do not set forth clear requirements regarding the exact form, style and timing of the disclosures, which makes

© Springer International Publishing AG 2017
S. Feuerriegel and D. Neumann (Eds.): FinanceCom 2016, LNBIP 276, pp. 62–74, 2017.
DOI: 10.1007/978-3-319-52764-2_5

the *immediate* publication a subjective rule open to some degree of flexibility (e.g. with regard to framing and publication time of the news) within the given regulatory constraints. Hence, we want to understand how the tone, or sentiment, of such disclosures relates to the abnormal stock price returns and publication strategies of stock-listed companies.

Methodologically, we first evaluate which are the main factors affecting the abnormal return. Then we inspect how sentiment metrics influence publication frequency. Finally, we analyze the relationship between the sentiment and the publication time by creating three dependent dummy variables for the different publication periods: *before trading, intraday trading* and *after trading*. We find that more than two-thirds of the messages are published intraday. We observe a positive correlation between the sentiment and abnormal return. News, which are released after trading hours, are significantly longer and marked by a negative effect of sentiment on abnormal returns.

In the following, Sect. 2 reviews related literature and presents our hypotheses. Section 3 introduces sentiment analysis and Sect. 4 describes our dataset, including descriptive statistics. In Sect. 5, we present our results and their evaluation, which we discuss in Sect. 6. Section 7 concludes with a summary.

2   Related Work and Research Assumptions

In recent years, various research has investigated the disclosure content bias. This research stream studies when and how companies publish new information, which is relevant for stock prices.

Bloomfield [22] and Li [14] conclude that companies sugarcoat negative news by making the content harder to understand and by avoiding the use of negative words. However, the company may be sued if the delay of the publication is long enough and the managers' reputation can suffer in the case of a failure to disclose news in a timely manner. Using a survey of 401 executives, Graham et al. [10] disentangle the content of positive and negative news. This study reveals that delaying bad news leaves more time for analysis regarding the likely perception, fine-tuning of the wording and mitigating the severity of the content with the right words. However, companies can enhance their reputation among analysts and investors by disclosing the bad news faster to assure transparency and accountability to their stakeholders. Kothari et al. [23] use dividend changes and managerial earnings forecasts to study whether managers withhold bad news. They find that on average, good news tends to be released quickly into the market, while managers withhold much of the bad news until its publication becomes unavoidable.

We acknowledge the vital impact of the timing of financial news publication and its influence the investor behavior and market returns. Therefore, this paper studies the relationship between the news sentiment, i.e. the positivity or negativity of the language used in a stock market disclosure, and the chosen publication time. We expect news related to factually more negative events (and thus with a lower sentiment) to be published with a delay, the potentially damaging information being deferred to a time when markets cannot immediately react, i.e. most often before or after trading hours. As such news contains factually negative content, we assume that it is characterized by a higher level of negative sentiment. In addition, we anticipate that negative disclosures will contain more words in comparison to the positive news; thus we analyze how the news length differs between these two categories.

3   News Sentiment Analysis

Sentiment analysis refers to analytical methods that measure the positivity or negativity of the content of textual data sources [24]. In this way, sentiment analysis can shed light on how human agents process and respond to the textual content of news.

Before investigating the differential information processing of news sentiment, we need to pre-process our news corpus according to the following steps:

1. *Tokenization:* tokenization splits a text into single word tokens [8,18].
2. *Negation inversion:* then, we account for negations using a rule-based approach to detect negation scopes and invert the meaning accordingly [5,20].
3. *Stop word removal:* in a next step, we remove so-called stop words, which are words without relevance, such as articles and pronouns [13].
4. *Stemming:* finally, we perform stemming in order to truncate all inflected words to their stems, using the Porter stemming algorithm.

After completing the pre-processing, we can study the influence of news sentiment on financial markets. We pursue a dictionary-based approach, in which a specific dictionary defines which words have a positive or negative connotation in a certain context (in our case the financial market). Here, we choose the Net-Optimism metric [6] combined with Henry's Finance-Specific Dictionary [9], since this is a common sentiment approach that leads to a robust relationship. The Net-Optimism metric $S(A)$ is given by the difference between the count of positive $W_{pos}(A)$ and negative $W_{neg}(A)$ words divided by the total count of words $W_{tot}(A)$ of an announcement A. We introduce the variable denoting news sentiment formally by

$$S(A) = \frac{W_{pos}(A) - W_{neg}(A)}{W_{tot}(A)} \in [-1, +1]. \tag{1}$$

All sentiment values are standardized, i.e. with a standard deviation of 1 and a mean of 0, in order to facilitate the understanding and interpretability of the results.

4   Dataset and Descriptive Statistics

4.1   Regulatory Filings from the German CDAX

We use a dataset that contains regulatory filings, i.e. mandatory news announcements (also referred to as ad hoc filings), published directly by stock-listed companies. Due to German stock market law, these news announcements need to be published by qualified issuers to ensure that all market participants have the same unrestricted access to information. In the German stock market, the leading news issuing service is DGAP.[1] Thus such ad hoc filings are released on DGAP, among other services. This regulation regarding data collection allows us to analyze the writing style of the original news items published by the companies under scrutiny.

As the law dictates that the information contained in these regulatory filings has to be stock-relevant and may not have been previously published through any other information channel, this dataset is highly suited to the study of information processing in financial markets. In other words, the information in the regulatory filings is novel and stock-price-relevant at the same time. This suggests that we do not observe a reverse causality of abnormal returns on the linguistic tone or information in a filing and likewise indicates that this dataset does not face an endogeneity problem with regard to the news content. However, we do not directly evaluate the causality in the examined relationships.

Our dataset contains 14,427 messages released on the DGAP platform between May 01, 2004 and June 27, 2011 by companies listed on the German CDAX index. Figure 1 shows the distribution of the news published within each hour. As expected, most of the messages are published between 07:00 and 09:00 a.m., shortly after trading hours start, and the frequency of news publication drops dramatically during the night.

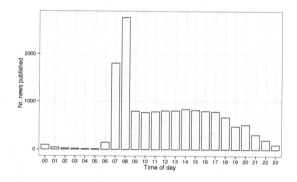

Fig. 1. Number of published news items by hour

4.2   Announcements Tone Metrics

Our dataset contains the following tone metrics: the *sentiment* (the net sentiment variable as defined in Sect. 3), the *positivity* and the *negativity*.

Positivity and *negativity* are represented by the share of positive and negative words, respectively. We compute these partial sentiment scores with Henry's Finance dictionary, which is frequently used in finance research [15].

---

[1] Deutsche Gesellschaft für Ad-hoc Publizität (DGAP). http://www.dgap.de/.

The sentiment variable thus captures the perceived positivity or negativity of a message. The sentiment metric is based on the difference between the *"observed relative frequency of a word in a positive or negative message and the expected conditional probability for a word to be in a positive or negative message based on its general occurrence in the overall corpus"* [16]. We focus our attention mainly on this sentiment metric, since it incorporates and summarizes the net impact of language.

4.3   Descriptive Statistics and Correlation Analysis

Table 1 shows the descriptive statistics of our relevant variables. The *message length* is a straightforward measure and it consists of the total number of words for each of the news items. On average, the news releases consist of around 323 words each, with a standard deviation of 321, which shows that the messages vary greatly in length. The shortest message contains only 3 words, while the longest message contains 11,451 words. The daily abnormal return measure reflects the deviation of the observed return from the expected return [12,17]. A more negative abnormal return indicates that investors evaluate the information in a new disclosure as more negative than expected. The converse also holds true, with investors evidently judging the information to be more positive than expected in the case of a positive abnormal return.

Table 1. Descriptive statistics

Statistic	Mean	St. Dev.	Min	Max	N
Sentiment	−0.00	1.00	−7.62	33.38	14,427
Positivity	0.00	1.00	−0.79	14.60	14,427
Negativity	0.00	1.00	−0.48	19.67	14,427
Message Length	323.03	321.81	3	11,451	14,427
Return	0.82	9.03	−81.07	300.00	14,427
Abnormal Return	0.75	9.12	−80.02	311.57	14,427
Log Return	0.38	8.56	−161.05	141.48	14,427

The correlation matrix from Table 2 provides a good understanding of the relation among the variables and gives insights into the variable selection and how to check for possible multicollinearity issues.

We see that positivity and negativity are positively correlated with message length. The same matrix shows the relation between different market controls, a considerable correlation between abnormal return, market volume and the price-to-book ratio. We address this issue and check for potential multicollinearity by computing Variance Inflation Factors (VIF) for our regressions. VIF values for all our variables of interest are smaller than 1.9. Thus, no multicollinearity concerns exist.

Table 2. Correlation matrix

	Sentiment	Positivity	Negativity	Message length	Hour	AR	ln(MV)	ln(PtB)
Sentiment	1							
Positivity	0.153	1						
Negativity	−0.201	0.462	1					
Message length	−0.129	0.651	0.655	1				
Hour	−0.077	−0.240	−0.107	−0.137	1			
AR	0.070	0.015	−0.063	0.0004	−0.009	1		
ln(MV)	0.058	0.009	−0.058	0.001	−0.018	0.818	1	
ln(PtB)	0.067	0.015	−0.064	0.001	−0.012	0.949	0.796	1
Alpha	0.014	0.013	−0.009	0.020	−0.016	−0.156	−0.014	−0.020

4.4 Message Length and Sentiment

Our hypothesis is that negative events require more time for analysis, which can then be used to downplay the full magnitude of the bad news with an excess of words. Such a relativization of negative events can be achieved through additional and unnecessary explanation of the events. We compare the average length of positive and negative messages. The means are computed for all 14,427 news in the sample. The null hypothesis for the one–sided test is: H_0: $|Positivity| \leq |Negativity|$.

Messages with a negative sentiment contain an average of 328 words, while the positive counterparts contain 318 words, a decrease of 3.05%. The difference is statistically significant at the 5% level. This result provides a first insight into the differing structure of the messages depending on their return direction.

4.5 Analysis of Sentiment for Different Trading Windows

We are interested in seeing the aggregate effect of news publication time during *intraday, before trading* and *after trading* hours and perform a comparison between them. The *intraday* group includes news published between 09:00 and 17:30, *after trading* – between 17:30 and 00:00, while *before trading* comprises messages published from midnight until 09:00. We use the working hours as stated on the official website of Xetra[2], since it is a common trading platform.

As Table 3 reveals, the abnormal returns are highest when companies release their ad hoc news intraday. Ad hoc news disclosed before trading hours are characterized by a more positive sentiment value, but are linked to the lowest abnormal returns across the three different observation periods. The messages published after closing hours tend to have a lower sentiment value compared to those published intraday. As expected, 67% of the disclosures are published during the working day, while the other two categories share 15% and 18% of the total news. In summary, these descriptive statistics are an early indication that companies tend to publish ad hoc news containing less favorable factual

---

[2] http://www.xetra.com/xetra-de/handel/handelsinformationen/handelszeiten/.

Table 3. Tonality value before trading, intraday and after trading

	Time frame	Sentiment	Abnormal return	N
1	Before trading	0.1716	0.1321	2,144
2	Intraday	−0.0204	0.4910	9,631
3	After trading	−0.0648	0.1977	2,652

information (and thus leading to lower abnormal returns) before and after trading. In particular, we observe a higher sentiment score for news released before trading.

5   Results

5.1   Analytical Framework

This research sheds light on how the language tone, or sentiment, relates to the abnormal return and the time at which stock-listed companies publish new disclosures. First, we will look at the main factors that affect the abnormal return, and then inspect how publication frequency is influenced by our sentiment metrics with

$$y_t = \beta X_t + \gamma \, C_t + \epsilon_t, \tag{2}$$

where y is the dependent variable, X is a vector of independent variables used subsequently: the sentiment metrics and the message length, and C includes a set of control variables for the market volume, price-to-book values, cumulative abnormal returns and active returns on the investment.

Finally, we will analyze the impact of publication time on sentiment values by creating three dependent dummy variables for the following categories: before trading, intraday and after trading. We employ logit regressions in order to compute the probability that a message is published at a certain interval given the sentiment values with

$$P(y_t = 1|x_t) = \Lambda(\beta X_t + \gamma \, C_t + \epsilon_t) \tag{3}$$

5.2   Influencing Factors of Abnormal Returns

We will use a set of linear regressions with different sentiment metrics in order to understand the relation between the abnormal return and the sentiment values. The dependent variable for Eq. 2 is the abnormal return as defined by [21], to reflect the unexpected stock price reaction on the day the news is emitted.

The stock prices are positively skewed, as a non-linear relationship exists between the independent and dependent variables. In order to improve our measure, we will use abnormal log-returns. Taking the log of the dependent variables normalizes the distribution [4].

Table 4. Comparison of sentiment measures for prediction of abnormal log return

	(1)	(2)	(3)	(4)	(5)
Tonality	0.187***				0.159***
	(0.050)				(0.050)
Negativity		−0.075**			−0.118**
		(0.030)			(0.051)
Positivity			0.087***		0.081***
			(0.025)		(0.026)
Message length				0.00003	0.0002
				(0.0001)	(0.0001)
ln(Market volume)	12.819	12.827	12.847	12.842	12.805
	(9.219)	(9.241)	(9.243)	(9.246)	(9.209)
ln(Price to book)	71.656***	71.734***	71.754***	71.772***	71.594***
	(11.674)	(11.690)	(11.685)	(11.689)	(11.676)
Alpha	−0.942***	−0.940***	−0.942***	−0.940***	−0.946***
	(0.055)	(0.055)	(0.055)	(0.055)	(0.055)
CAR	0.004	0.004	0.004	0.004	0.004
	(0.010)	(0.010)	(0.010)	(0.010)	(0.010)
Constant	−0.265***	−0.265***	−0.266***	−0.277***	−0.320***
	(0.037)	(0.037)	(0.037)	(0.049)	(0.051)
Observations	13,914	13,914	13,914	13,914	13,914
R^2	0.836	0.836	0.836	0.835	0.836
Adjusted R^2	0.836	0.835	0.835	0.835	0.836

Notes: ***, **, and * denote significance at the 1, 5, and 10 percent level.
The robust standard errors are clustered at firm level.

Table 4 reports the results obtained when the abnormal log-return is the dependent variable. We closely follow Tetlock [25] in choosing control variables, which originate from the Fama-French model and its extensions [3, 7, 11]. Column (5) reveals the results of our regression when including all sentiment measures. The effect directions of all sentiment measures are as expected and statistically significant. The sentiment variable has a coefficient size of 0.159 (significant at the 1% level), the negativity variable has a coefficient size of −0.118 (significant at the 5% level) and the positivity variable has a coefficient size of 0.081 (significant at the 1% level). The message length is not statistically significant.

The effect size of negativity on the abnormal log-return is larger than the effect size of positivity. Thus investors react more strongly to a shift towards negative rather than positive sentiment. This is consistent with Tetlock et al. [24], who finds that negativity has a stronger correlation with stock returns than positivity.

Table 5. OLS regressions for prediction of publication frequency

	(1)	(2)	(3)	(4)	(5)
Sentiment	−1.543*				−1.414**
	(0.833)				(0.656)
Negativity		1.041			0.576
		(0.640)			(0.765)
Positivity			−0.171		−0.124
			(0.433)		(0.566)
ALR				0.013	0.056
				(0.035)	(0.041)
ln(Market volume)	0.853	0.244	0.027	−0.498	−1.357
	(2.323)	(1.816)	(1.697)	(2.248)	(1.960)
ln(Price to book)	−3.681**	−3.417**	−3.719**	−4.076**	−5.220**
	(1.820)	(1.588)	(1.702)	(1.762)	(2.365)
Alpha	7.178***	7.070***	7.026***	7.028***	7.207***
	(1.122)	(1.065)	(1.051)	(1.046)	(1.135)
Observations	2,055	2,055	2,055	2,055	2,055
R^2	0.001	0.0002	0.0001	0.0001	0.001
Adjusted R^2	0.000	10.000	10.000	0.000	0.000

Notes: ***, **, and * denote significance at the 1, 5, and 10 percent level. The robust standard errors are clustered at firm level.

5.3   News Sentiment and Publication Frequency

Next, we want to study whether news sentiment and the frequency with which a company publishes ad hoc disclosures are related. We use OLS regression to estimate the model, but we change the data structure, the level of observation and both the dependent and explanatory variables:

$$y_{it} = \beta X_{it} + \gamma\, C_t + \epsilon_{it}, \tag{4}$$

where i denotes company code and t is the year of the publication. We have 492 firms and 8 unknown values for the companies in the current dataset and we span the period of May 2004–June 2011. *Publication frequency* is our dependent variable and it is defined as the number of news-items-per-year released by every company. The vector C contains additional covariates, namely abnormal log return, market volume, price-to-book and alpha value.

The results in Table 5 suggest that companies with a more positive tone publish fewer ad hoc announcements. In other words, a one standard deviation increase in the tone leads, on average, to −1.414 fewer announcements per year, significant at the 5% level. Interestingly, companies with a higher price-to-book value, i.e. those which have a rather high valuation, communicate less often via announcements, significant at the 5%level. While we find significant relationships, this model has to be treated with particular caution, as the R^2 leans towards 0. Hence, the model does only poorly explain the dependent variable.

5.4   News Sentiment and Publication Timing

After uncovering the relationship between the tone of the news and the stock performance, as well as the publication frequency, we examine the daily timing of disclosures. We use the econometric model from Eq. 2, in which the dependent variables are *before trading*, *intraday* and *after trading*. They are indicator variables equal to one if the news was emitted during that time and zero otherwise. The vector of control variables C comprises market volume, price-to-book and abnormal log return. We exclude the cumulative abnormal return and the market alpha from our analysis, since their results were not statistically significant and we prefer to have a more parsimonious model. The results are presented in Table 6.

Table 6. Logit regressions for different publication time intervals

	Before Trading	Intraday	After Trading
	(1)	(2)	(3)
Sentiment	0.131***	−0.040**	0.006
	(0.025)	(0.019)	(0.024)
Negativity	0.063**	−0.026	−0.010
	(0.028)	(0.024)	(0.034)
Positivity	0.588***	−0.212***	−0.468***
	(0.029)	(0.025)	(0.042)
Message length	−0.0003***	0.0001	0.0003***
	(0.0001)	(0.0001)	(0.0001)
ln(Market volume)	−0.082	0.480	−0.876
	(0.473)	(0.480)	(0.868)
ln(Price to book)	0.719	−0.076	−0.498
	(0.603)	(0.551)	(0.850)
ln(Abnormal return)	−0.013**	0.002	0.011
	(0.006)	(0.005)	(0.008)
Constant	−1.735***	0.670***	−1.641***s
	(0.044)	(0.034)	(0.044)
Observatiosns	13,949	13,949	13,949
Log Likelihood	−5,499	−8,796	-6,535
AIC	11,015	17,609	13,087

Notes: ***, **, and * denote significance at the 0.1, 1, and 5 percent level. The robust standard errors are clustered at firm level.

News released before trading are significantly more likely to have a more positive sentiment and a higher degree of positivity. The messages published before

09:00 are shorter at statistically significant levels. Disclosures published before trading are also characterized by a lower degree of negativity (significant at the 1% level). This might suggest that negative news from the previous day that were not all published before midnight and are actually partly only released before the trading hours, so that the releasing company has enough time to prettify them. This is in line with the findings of Graham et al. [10], who conclude that delaying the release of bad news leaves more time for analysis and communicating the content with the right words, i.e. a better tone. However, the abnormal log return shows a negative relation with the news released before trading, which might suggest that prettifying the news is not effective and investors are hardly tricked.

On average, intraday releases have lower sentiment and are inversely related to positivity. Such messages seem to appear shortly after the events in question and with a lower sentiment. There is no statistically significant relation between the intraday publication time and abnormal returns.

In the after trading period, between 18:00 and midnight, we observe a reduced usage of language positivity and an increased text length (both statistically significant at the 0.1% level). These findings might indicate that companies are using more words to describe certain less advantageous facts, which, on average, are less positive and thus less suitable for the deployment of positive language.

Previous research has investigated the motivation for publishing bad news in the evening. Botosan and Plumlee [2] argue that publishing bad news in the evening could lead to decreased volatility, since investors have more time to analyze the news and reflect on its content. Also, disclosing bad news on the same day decreases the probability of a lawsuit and can increase the company's reputation in recognition of its high transparency and accurate reporting [10].

6   Discussion

As our results reveal, disclosures, which are published before the markets open have a higher sentiment than those disclosures which are published during the day. At the same time, the abnormal returns are higher following a publication of a new disclosure during the day (0.491 percent) than before trading (0.132 percent) or after trading (0.198 percent). The publication of disclosures intraday leads to a more positive market reaction, although their content comes at a lower sentiment on average. This suggests that companies tend to phrase information in pre-trading disclosures in a more positive manner than in their intraday disclosures. This might be due to a strategic effort by stock-listed companies to prettify the language in order to soften the blow of negative information. However, with the given data, we cannot further validate or substantiate this finding, as we cannot make any judgment about the existence and direction of causality between the informational or sentiment content and the disclosure timing.

7 Conclusion

We investigate the role of positive and negative words, as well as the overall sentiment, in financial disclosures published by stock-listed companies. We measure the different sentiment measures by relying on a dictionary-based sentiment approach. This paper focuses on how the different sentiment variables relate to the stock performance, disclosure frequency and the timing of mandatory financial disclosures, German ad hoc announcements.

We find that negative messages are slightly longer in comparison to positive ones. The difference is statistically significant at the 5% level. Our OLS regressions show that overall sentiment has the strongest relation to the abnormal log return. Negativity and positivity show the expected correlation direction relative to the stock performance. The former decreases the abnormal log return, while positivity has a positive effect on abnormal returns. In addition, we observe larger abnormal returns following disclosures published during trading hours, despite the fact that they have a lower sentiment on average than those disclosures, which are published before or after trading.

This study cannot conclude that delaying news is related to higher stock returns, but we show that sentiment is an important indicator for the abnormal log return. One avenue for future research would be to employ quantile regressions in order to study potentially heterogeneous effects on different parts of the distribution.

References

1. Bundesministeriums der Justiz und für Verbraucherschutz: Gesetz über den Wertpapierhandel, 6 July 1994. https://www.gesetze-im-internet.de/wphg/
2. Botosan, C.A., Plumlee, M.A.: A re-examination of disclosure level and the expected cost of equity capital. J. Account. Res. **40**, 21–40 (2002)
3. Chan, L.K., Jegadeesh, N., Lakonishok, J.: Momentum Strategies. J. Finance **51**(5), 1681–1713 (1996)
4. Brooks, C.: Introductory Econometrics for Finance. Cambridge University Press, Cambridge (2014)
5. Dadvar, M., Hauff, C., de Jong, F.: Scope of negation detection in sentiment analysis. In: Proceedings of the Dutch-Belgian Information Retrieval Workshop (DIR 2011), Amsterdam, Netherlands, pp. 16–20 (2011)
6. Demers, E.A., Vega, C.: Soft information in earnings announcements: news or noise? SSRN Electron. J. (INSEAD Working Paper No. 2010/33/AC) (2010)
7. Fama, E.F., French, K.R.: Common risk factors in the returns on stocks and bonds. J. Financ. Econ. **33**(1), 3–56 (1993)
8. Feuerriegel, S., Neumann, D.: News or noise? how news drives commodity prices. In: Proceedings of the 34th International Conference on Information Systems (ICIS 2013). Association for Information Systems (2013)
9. Henry, E.: Are investors influenced by how earnings press releases are written? J. Bus. Commun. **45**(4), 363–407 (2008)
10. Graham, J., Harvey, C., Rajgopal, S.: The economic implications of corporate financial reporting. J. Account. Econ. **40**, 3–73 (2005)

11. Jegadeesh, N., Titman, S.: Returns to buying winners and selling losers: implications for stock market efficiency. J. Finance **48**(1), 65–91 (1993)
12. Konchitchki, Y., O'Leary, D.E.: Event study methodologies in information systems research. Int. J. Account. Inf. Syst. **12**(2), 99–115 (2011)
13. Lewis, D., Yang, Y., Rose, T., Li, F.: RCV1: a new benchmark collection for text categorization research. J. Mach. Learn. Res. **5**, 361–397 (2004)
14. Li, F.: Annual report readability, current earnings, and earnings persistence. J. Account. Econ. **45**(2–3), 221–247 (2008)
15. Loughran, T., McDonald, B.: Textual analysis in accounting and finance: a survey. J. Account. Res. **54**(4), 1187–1230 (2016)
16. Liebmann, M., Hagenau, M., Häussler, M., Neumann, D.: Effects behind words: quantifying qualitative information in corporate announcements. In: 33rd International Conference on Information Systems (ICIS 2012) (2012)
17. MacKinlay, A.C.: Event studies in economics and finance. J. Econ. Lit. **35**(1), 13–39 (1997)
18. Manning, C.D., Schütze, H.: Foundations of Statistical Natural Language Processing. MIT Press, Cambridge (1999)
19. Nicholas, C.: Barberis: thirty years of prospect theory in economics: a review and assessment. J. Econ. Perspect. **27**(1), 173–196 (2013)
20. Pröllochs, N., Feuerriegel, S., Neumann, D.: Enhancing sentiment analysis of financial news by detecting negation scopes. In: 48th Hawaii International Conference on System Sciences (HICSS). IEEE Computer Society (2015)
21. Pröllochs, N., Feuerriegel, S., Neumann, D.: Generating domain-specific dictionaries using bayesian learning. In: 23rd European Conference on Information Systems (ECIS 2015), Münster, Germany (2015)
22. Bloomfield, R.J.: The incomplete revelation hypothesis and financial reporting. Account. Horiz. **16**, 233–243 (2002)
23. Kotari, S.P., Shu, S., Wysocki, P.: Do managers withhold bad news? J. Account. Res. **47**(1), 241–276 (2009)
24. Tetlock, P.C.: Giving content to investor sentiment: the role of media in the stock market. J. Finance **62**(3), 1139–1168 (2007)
25. Tetlock, P.C., Saar-Tsechansky, M., Macskassy, S.: More than words: quantifying language to measure firms' fundamentals. J. Finance **63**(3), 1437–1467 (2008)
26. Thaler, R.H.: Advances in Behavioral Finance. Roundtable Series in Behavioral Economics, vol. 2. Princeton University Press, Princeton (2005)

FinTech Transformation: How IT-Enabled Innovations Shape the Financial Sector

Liudmila Zavolokina[✉], Mateusz Dolata, and Gerhard Schwabe

Institute of Informatics, University of Zurich,
Binzmühlestrasse 14, 8050 Zurich, Switzerland
zavolokina@ifi.uzh.ch

Abstract. FinTech, the phenomenon which spans over the areas of information technologies and financial innovation, is currently on the rise and is gaining more and more attention from practitioners, investors and researchers. FinTech is broadly discussed by the media, which constitutes its understanding and represents social opinion, however, this perception of FinTech should be supported by empirical evidences. Therefore, we examine five Swiss FinTech companies through the lens of the conceptual framework of understanding of FinTech and its dimensions and, by doing so, analyze the nature of FinTech innovations. Thereby, we extend the understanding of FinTech and provide a fruitful soil for further research in this area.

Keywords: FinTech · Financial innovation · Digital innovation · Disruption · Business model · Blockchain

1 Introduction

Classical banking has undergone considerable changes in the last decades and currently is facing the new era of digitalization. Digital technologies interfere almost each part of the banking business: from private banking to investment banking, from treasury to risk management. However, the digital technologies create value not only in the banking sector, but rather stretch along all the possible fields of financial sector. FinTech, which originates from the intersection of technology and finance, represents this intrusion of the digital technologies into the businesses enabling the latter to innovate. Although, the digital technology itself does not provide any advantage alone, but rather, combined with other available resources (e.g., organizational, environmental, strategical), it can create business value.

In this paper, based on the example of a few FinTech companies, we examine three steps of FinTech transformation and, by doing so, validate the conceptual framework of FinTech, proposed in [1]. This framework represents and synthesizes the social opinion on the understanding of FinTech phenomenon and, therefore, gives us the starting point to explore it empirically. In particular, we analyze five important and quite successful players on the Swiss Fintech market. First, we focus on the triggers which motivate FinTech innovation in financial services, namely the combination of a technology, an organization, and investments. Second, we look closer at the transformation which this innovation performs. Third, we explore how such transformation affects FinTech

S. Feuerriegel and D. Neumann (Eds.): FinanceCom 2016, LNBIP 276, pp. 75–88, 2017.
DOI: 10.1007/978-3-319-52764-2_6

services, products, business processes and business models. To do that, we analyze different kinds of sources (official websites of the companies, industrial reports and studies, newspaper articles, presentations and researchers' notes made during conferences for practitioners) and extract the information to each of the dimensions, discussed above.

The contribution of this paper is mainly targeted at practitioners in the field of digital and financial innovations, who would like to examine the empirical evidences of FinTech transformation on the example of the Swiss financial market. However, it also brings value to the researchers, who lack knowledge on the area of FinTech. To do that, we empirically validate the conceptual framework of understanding of FinTech, presented in [1], and discuss it from the perspective of practical applications. Here, we explore, if and how the building blocks, constituting the framework, are reflected in practice. Additionally, we argue if the conceptual framework can be used for identifying successful FinTech companies and projects. By doing so, we extend the scientific literature on FinTech and, therefore, contribute to innovation literature.

The remainder of the paper is organized as follows. In the related work we present the conceptual framework of FinTech from the perception of the popular media, used for the analysis, and discuss its dimensions (input, mechanisms and output) in more detail. The section "Snapshots of the Studied Companies" introduces the analyzed Swiss FinTech companies from 5 different areas of FinTech and syntheses the information in relation to the framework. Thereafter, we come to the findings of this study, where we discuss standing out features of the companies, their challenges and opportunities and role on the Swiss FinTech arena. This brings us to the conclusion of the paper, which admits the limitations of this study, but also highlights further research possibilities.

2 Related Work

The objective of this work is to validate the conceptual framework of understanding of FinTech on the practical examples of the five Swiss companies, operating in five different fields within FinTech. Therefore, in this chapter we will introduce the reader to the notion "FinTech", existing literature and the framework, which presents how FinTech is perceived by the popular media, and present the FinTech areas, where the studied Swiss companies come from.

Due to its novelty, the hyped phenomenon of FinTech itself is rather underrepresented in the Information Systems (IS) research. A keyword search for "fintech" in the AIS Electronic Library yields 11 hits (last accessed on 01.09.2016), which focus not on the phenomenon itself, but rather mention it in the context of "financial technology" once or twice (except for the paper published in 2004, where the name "Fintech" was related to the company, which published the cited report). Although, the scientific definition, which common to be used in IS research, is still missing as well as the unified concept of FinTech phenomenon.

In order to shed the light onto the phenomenon, in [1] the authors analyze more than 800 articles coming from the leading newspaper outlets, which are a representative of socially constructed opinion, and make the following conclusions. According to this study, FinTech is perceived as the process of transformation which lies on the intersection of financial and digital innovations and triggered by three elements, namely the underlying technology, organizational body and incoming investment. Such triggers, or enablers, are the mechanism for fostering the FinTech innovation. By the transformation four functions are meant: creation/change/improvement, disruption, application of IT to finance, creation of competition. They represent the actual processes which leads to the third dimension – the outcomes of such transformation. The result of this transformation produces new services/products/processes/business models. Figure 1 illustrates the conceptual framework. Further, we will introduce some of the concepts relating to the building blocks of the framework which may seem ambiguous and, therefore, mislead the reader. In our study we relate "create OR change OR improve" applied to the output dimension, namely services, products, processes and business models. We consider the disruption function of FinTech through the light of a disruptive technology, which is defined as "a technology that changes the bases of competition by changing the performance metrics along which firms compete" [2]. By the application of IT to finance we relate to the use of the "incoming" technologies (for example, blockchain) to the financial scenarios. Furthermore, the creation of competition is necessary for FinTech in a way, that innovating FinTech companies attract attention and trigger reaction among their competitors [3].

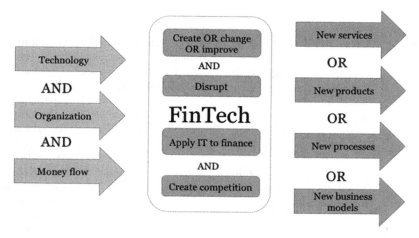

Fig. 1. Visual representation of the integral definition of FinTech [1]

However, the proposed framework does not reflect the actual behavior of FinTech companies and thus needs to be validated in practice or even extended. The reason for that may come from the nature of the evolvement of the framework, as it mainly presents the understanding of the notion "FinTech" through the popular media and is rather a reflection of the actual phenomenon [1]. Therefore, it is important to deepen this research and examine examples of FinTech transformation in "the real world".

3   Snapshots of the Studied Companies

In this section we provide an overview of five Swiss FinTech companies and explore them through the lens of the conceptual framework, discussed above. As far as the framework suggests three dimensions of the FinTech transformation – namely the input, mechanisms and output – we look at the companies with the special focus on these three dimensions. The companies come from five different areas of FinTech in the order they appear in this section: peer-to-peer lending, payments, insurance, finance management and blockchain. Tables 1 and 2 synthesize the reflection of the conceptual framework in the studied companies and are followed by the extensive description.

Table 1. Five Swiss FinTech companies through the prism of the conceptual framework (input)

	Input					Money flow
	Technology	Organization				
		Year	Location	Market	Type	
CreditGate24	- Crowdlending platform - Automated credit scoring system	2014	Zurich	B2B National	Independent company	Not available
Twint	Digital wallet in an app	2014	Bern	B2B & B2C National	Subsidiary	Funded by PostFinance
Knip	Digital insurance broker in an app	2013	Zurich	B2B & B2C International	Independent company	- Funded by venture capitalists - The most funded FinTech company in Switzerland
Contovista	Software for personal finance management	2013	Zurich	B2B National	Startup	Funded by large financial service and software providers
Ethereum	Blockchain platform	2015	Zug	Non-profit	Foundation	Funded through crowdfunding campaign

3.1   CreditGate24

CreditGate24, founded in 2014, connects borrowers with private and institutional investors through its highly automated direct-lending online platform [4]. The company operates on the national level.

Table 2. Five Swiss FinTech companies through the prism of the conceptual framework (mechanisms and output)

	Mechanisms		Application of IT in finance	Creation of competition	Output
	Creation/change/improvement	Disruption			
CreditGate24	- Low costs - Attractive interest rates - Flexible settlement process (online & quick)	Replacement of an intermediary	Lending business	In the top 11 Swiss FinTech platforms	Service process
Twint	*For business customers:* - Price of transactions - Mobile marketing channel - Easy and low-cost setup *For private customers:* - Easy and quick cashless payments - Expenses management - Loyalty programs - No additional costs	Replacement of credit card payments	Payments	One of the most important players in Switzerland	Service
Knip	*For business customers:* - New sales channel *For private customers:* - Free service - Overview of insurances in one app - Free advisory service (per phone, email or online chat)	Replacement of insurance brokers	Insurance management	In the top 100 FinTech companies	Service
Contovista	Saving costs	Not disruptive	Personal finance management	In the top 100 Swiss FinTech startups	Product
Ethereum	- Secure and networked computing - Smart contracts can regulate any relationship which can be digitized	Replacement of an intermediary	Smart contracts can regulate any relationship which can be digitized, going beyond financial area	- The largest competitor of BitCoin - Does not compete but provides the technology which creates competition among its users	Business model

Technology. The peer-to-peer lending platform allows to perform credit check based on classic credit assessment methods, big data analytics, the insurance and the solidarity arrangements [4].

Organization. CreditGate24 (Schweiz) AG is an independent Swiss company. The company is regulated by VQF (The Financial Services Standards Association), what indicates that the services are compliant with regulatory standards [4].

Money flow. Information on the investments, pouring into the company, is not available in the open sources, however, it is known that the company is not considered to be a startup, but works independently.

Creation/change/improvement. The company provides low-cost services and transparent fees, attractive interest rates (credit history and reliability of the borrower influences his/her rating; the better the rating is, the more attractive the interest rates are). Furthermore, the settlement process occurs online, flexibly and quickly, as both lenders and borrowers do not have to come to the branch offices, but communicate through the platform [5]. Therefore, we conclude that the company improve the costs of the service for the customers as well as it changes the process of handling the settlements.

Disruption. The disruptive function of the CreditGate24 innovation is presented by replacing the intermediary like a bank as a lender or investment advisor and bringing borrowers and investors together. Moreover, the platform allows to avoid going to the bank's office to arrange the credit settlement, but to perform all the activities online. However, Swiss banks are looking for space for cooperation. For example, Hypothekarbank Lenzburg, which before has not offered private loans, announced its cooperation with CreditGate24 in order to extend their offer [6].

Application of IT in finance. The company offers an online platform, which means there is no office, where the clients would come to, but on top of that the credit scoring system, integrated into the platform, is automated and uses methods from big data analytics. For example, it utilizes the data from the social networks to verify the information about borrowers and investors.

Creation of Competition. CreditGate24 is one of more than 40 peer-to-peer lending platforms in Switzerland and is considered to be one of Europe's top 11 the most important platforms [7]. The management's ambitions are to become the largest lending platform in Switzerland [5], what means struggling with high competition on the market of crowdlending platforms. Among current competitors in Switzerland are Cashare – another crowdfunding platform – or Migros Bank as classic financial service provider with low interest rates on its loans [8].

Output. CreditGate24 is not a bank but a broker, therefore, the output of its business is providing the service which is used by both parties – investors and borrowers. Moreover, the company also changes the process of lending and investing money – switching from classical visit of an office in a bank to automatized online settlement. However, the provided platform (and the lending business itself) is strongly dependent on the matter of trust, this is where the banks bring still more advantage [8].

3.2   Twint

Twint AG, founded in 2014, is aimed at providing digital wallet service for use in Switzerland [9]. Twint targets both private and business customers on a national level.

Technology. For private clients Twint provides a mobile application, which allows its users to pay cashless in shops (both online and offline) and to transfer money between Twint-users [9]. For business clients Twint provides "Twint Beacon", which allows to settle the payment via Bluetooth (the beacon is a Bluetooth sender, which allows to pay cashless at the cashier; the beacon sends a number, which identifies the shop and the cashier, at which a client is standing [10]), and an application for business clients. Furthermore, business clients can start their marketing campaign, in which they can issue digital coupons and stamp cards [9].

Organization. Twint AG is a 100% subsidiary of PostFinance, a large Swiss financial service provider. The CEO of Twint is a member of PostFinance Executive Committee, whereas the chairman of Twint is the PostFinance CEO [10].

Money flow. Though the information on the investments is not available online, but as far as the company fully belongs to the PostFinance, we can assume, that the company was fully funded by its elder brother PostFinance, which in its turn is a subsidiary of the Swiss Post.

Creation/change/improvement. According to [10], the advantage which Twint brings is twofold: for business customers and for private app-users. For business customers the improvements are the following: the price of a transaction is lower for all sales channels (web-payments, cashier, app), mobile marketing channel at hand, easy and low-cost set-up. For private customers: easy and quick payments with the app (no credit card needed), expenses management in the app (overview of expenditures), loyalty programs and offers from the partners (coupons, digital stamp cards), no additional costs.

Disruption. The disruptive character of the service, provided by Twint, is presented by replacing credit card payments with payments via smartphone, from the one hand. From the other hand, it allows business customers to process payments cheaper, faster and easier, but also to perform their marketing strategy using Twint.

Application of IT in finance. In case of Twint, the area of payments is exposed to the transformation, where the app for digital wallet is introduced. The payment can be settled through the "Twint Beacon" (which communicates with the cashier via Bluetooth), numeric token or QR-code.

Creation of Competition. Twint is one of the most important players in the area of digital payments in Switzerland (competing with Paymit, TawiPay, CashSentinel [11]). However, Twint claimed that the payment platform will be united with Paymit, integrating the best features from both into one solution [12]. On top of that, surprisingly, the PostFinance launches the competitor to their own product within the enterprise – an app which allows to pay using NFC-function [13].

Output. As an output, Twint creates a new service of payments for its private clients, but also opens a new channel for the marketing strategy for business clients.

3.3   Knip

Knip AG, founded in 2013, is a mobile insurance broker, which allows its users to compare tariffs and services from different companies [14]. Knip operates in Switzerland starting from 2013 and in Germany starting from 2015. Knip partners with insurance companies, whose offers are presented in the app.

Technology. Knip provides an app of the same name, which shows available insurance contracts, polices and offers. The technology is supported by human interaction in the way that insurance experts provides advice on security and tariffs online in chat or by phone [15]. The app is available for iOS- and Android users.

Organization. Knip AG currently has more than 100 employees, distributed in Europe (Switzerland, Germany and Serbia). The company is registered with IHK, the German Chamber of Industry & Commerce, and certified by TÜV, the German Technical Inspection Agency [16].

Money flow. As a startup, Knip was funded by venture capitalists from the U.S., Switzerland, Netherlands, Germany. By the end of the year 2015, it was announced that the company got the largest amount of investments on the FinTech arena in Switzerland [17].

Creation/change/improvement. The client gets an overview of his/her insurances, on top of that the client receives an advisory service for free, provided by "Knippers" through any channel (phone, email, online chat) [18]. Therefore, the service for the end customer is for free. Business customers (insurance companies, partnering with Knip) get one more sales channel in their turn. Therefore, the company creates the new service of accessing his/her insurance data and changes the channel for accessing this data.

Disruption. Though Knip is very successful among its competitors, its disruptive effect is doubtful, as the app is new, but its business model is not. The questions of transparency of the service and data security still remain open [19]. However, we should admit that the possibility to replace insurance brokers with an app exists.

Application of IT in finance. Although we consider Knip to belong to FinTech companies, it becomes more obvious that insurance startups are getting their own area of "tech" – InsurTech [20]. In case of Knip, the created app is applied to the area of insurance management.

Creation of Competition. In 2015 Knip got the 29th place in the FinTech 100 rating, created by KPMG and H2 Ventures, presenting the most successful and innovative startups in the area of FinTech. Concerning the local competition among startups, Knip remains a leader on the Swiss market in insurance management services, however, more and more startups appear trying to beat it (see Esurance, for example).

Output. Knip creates a service, which allows for cheaper (free, in fact) and easier (available through the app on a smartphone) service of insurance management.

3.4   Contovista

Contovista, founded in 2013, is a startup which created a software for personal finance management, which can be integrated into online banking [21]. The company targets B2B national market. The proposed solution has become known through the partnership with Zurcher Kantonalbank, the largest cantonal bank and the fourth largest bank in Switzerland [22].

Technology. The software, created by Contovista, is a personal finance manager, which automatically analyses and categorizes the expenses and income of a user. The software can be integrated into the solutions for online banking.

Organization. The company has 15 employees [23]. In the early 2016 the Aduno Group, Swiss expert in cashless payments, acquired 14 percent holding in Contovista and represented in Board of Directors [24].

Money flow. Contovista attracted considerable investments from the Aduno Group, large financial service provider [24], and the largest Swiss banking software providers, Avaloq and Finnova [21], which helped the company to boost.

Creation/change/improvement. Though the solution, created by Contovista, is not revolutionary, the benefit it brings to the customers is saving costs on in-house programming of the software [21]. Therefore, we conclude Contovista's solution as the creation of a new product.

Disruption. Although the software, proposed by Contovista, is not disruptive itself, the practice shows that only a few banks (UBS and PostFinance) developed such solution for their online banking service [21]. In this sense, we can conclude, that Contovista rather fills the gap, than disrupt the existing solutions.

Application of IT in finance. The application of IT in finance in case of Contovista is straight-forward: the provided software (IT) operates in the area of personal finance management and is implemented in the banking sector, which has direct relation to finance.

Creation of Competition. Contovista successfully competes among other FinTech startups: It entered 100 best Swiss FinTech startups, taking the 26th place [25]. However, the market of direct competitors with similar solutions is not diverse. The main competitor on the Swiss arena is Qontis, which also targets private clients [21].

Output. Contovista creates a product, the personal finance management software, which can be integrated in online banking service. This product is solely targeted on business customers, primarily banks.

3.5   Ethereum

Ethereum is a decentralized open-source platform that runs smart contracts, automated applications that run exactly as programmed on a custom built blockchain [26]. The Ethereum blockchain was launched in July 2015.

Technology. Smart contracts are programs that store their state on the public blockchain. These smart contracts (basically, lines of code) are written in Solidity, JavaScript-like programming language for developing smart contracts, which is compiled to bytecode, executable on Ethereum Virtual Machine [27]. Ether is a kind of fuel for operating the network, which clients pay to the machines to execute the transactions [28].

Organization. The platform is created by the non-profit Swiss-based Ethereum Foundation, whose mission is "to promote and support research, development and education to bring decentralized protocols and tools to the world that empower developers to produce next generation decentralized applications (dapps), and together build a more globally accessible, more free and more trustworthy Internet" [29].

Money flow. The Ethereum Foundation got funded (15 Million Swiss Francs) through the crowdfunding campaign in 2014 [29, 30] and became the second most successful crowdfunding project [31].

Creation/change/improvement. Ethereum enables secure and networked computing. It means there is no central authority, the network of computers does not require any trust from its participants, as the system regulates itself [30]. If to speak about applications of Ethereum in the financial area, Ethereum apps provide the possibility to execute financial transactions faster than in existing systems. Furthermore, smart contracts are well-suited for those kinds of transactions that involve interactions of counterparties (which in their turn require higher level of trust and lower risks, e.g., loan/debt payments). It means that the inter-banking activities can be improved the most [30], as the ones which require high level of trust by their nature and, therefore, the trustless architecture of the blockchain technology can be advantageous.

Disruption. Ethereum itself is not disruptive, however, the technology and the idea which run behind smart contracts, built on Ethereum, are. Money transfer without a bank, which coordinates and reduces risks of fraud; or bets without a betting shop – the removing of a central authority becomes possible with Ethereum [30].

Application of IT in finance. The applications of Ethereum platform go far beyond financial area, even though it currently remains the most attractive one (for example, crowdfunding campaigns run on Ethereum) [32]. Smart contracts may be used whenever their business logic allows to be transferred into code (existing examples include decentralized music streaming or observation of the voting [32]).

Creation of Competition. Ethereum is recognized to be the largest competitor with BitCoin, cryptocurrency running on blockchain. However, Ethereum is rather a creator of competition, than a competitor by itself. More and more startups are born on the idea of Ethereum and its smart contracts, more and more of existing ones try to utilize smart contracts for their business needs.

Output. Ethereum is not a service, but rather an enabler of a new business model and a process. There are startups utilizing the platform for creating new services and products, or using the technology in the existing businesses [33]. However, also IT giants IBM (bringing blockchain to Internet of Things) and Microsoft (with blockchain-based cloud service) approached Ethereum playing around with the technology and trying to integrate it in their businesses [30].

4 Findings

The purpose of this study is to examine the conceptual framework of the perception of FinTech, proposed by [1], from an empirical perspective. In this section we discuss obtained results, make conclusions and highlight the special outstanding features of the studied companies, observed in the data, in relation to the building blocks of the framework (input, mechanisms and output).

In this paper we studied five different, but quite successful FinTech players in Switzerland. The Swiss start-up and entrepreneurial scene is currently attracting more attention. The potential for entrepreneurship in the area of FinTech in Switzerland is high thanks to the considerable financial knowledge and experiences, however, there are some challenges to overcome at the regulatory and institutional levels. Furthermore, from the perspective of the investment flow Switzerland is often considered to be an attractive place, however, the analysis shows that the funding rather comes from the larger players on the arena than from external sources (like a crowdfunding projects). The considerable money flow from banks or large IT companies may be explained by their intention to cooperate with smaller FinTech startups, but not to compete with them and lose at the end of the game.

Considering the output dimension of FinTech, the tendency is to present oneself as providing services or business models. This goes in line with the whole turn of the economy from product-centric model to the service-logic. In our opinion, FinTech is a child of this re-orientation: for companies, which started their existence in the product-oriented time (banks still define, e.g., investment portfolios and their management as a product, which is sold by the frontline employees) such as PostFinance, doing service innovation can be a chance to stabilize their position in the new, service-oriented market.

Considering the mechanisms, involved in generation of FinTech innovation: we observe that the business offering, described in "change/creation/improvement" can be only seldom defined by a single item. On the contrary, each startup has a multifold contribution to the market, where some changes accompany creations (i.e., new offerings). While this is in line with the original formulation "Change OR Creation OR Improvement" (assuming that OR is an inclusive or), we see this field as extremely broad. Since the original framework was based on newspapers' articles, it shows how broadly the term was defined therein. However, here we can observe common patterns in cost reduction, increase of transparency, easiness and quickness of a service/process.

Considering the building block of the "technology" in the input dimension, we wish the original framework would make concrete suggestions on what technologies are involved in FinTech. What we observe in our analysis, is a tendency towards mobile devices on the one side (primarily for the consumer in a B2C scenario) and extensive platforms (including bitcoin for B2B scenario). This tendency opens up the questions, if this is really the direction FinTech will take in future or this is only so due to the focus of the studied companies on those two dynamically changing areas (which, of course, get reflected in the framework).

In general, we observe that the framework, proposed in [1], can be used to identify primary differences between startup companies in terms of organizational and business inputs; we, however, wish more concrete and precise definitions in the topic of technology and change/creation/improvement. However, these drawbacks reflect clearly the origin of the framework: while popular media is strong at identifying and describing business issues, and by discussing organizational topics, articles presented in there provide little insights on technologies used. Therefore, this dimension remains very general in the framework, proposed by [1], and clearly orients itself at presenting almost everything as "change OR creation OR improvement" as the newspapers thrive for sensational and attracting attention publications.

This study was a try to apply the proposed framework for extending the notion of FinTech by analyzing particular players – it was a successful try. We learned more about the service-oriented nature of FinTech and about the fact that, at least in Switzerland, successful innovations in this regard are driven by larger players who do not want to miss the opportunity to innovate and turn towards new services. However, we claim that a predefined classification schema in the technology as well as some help to better describe the creation or change or improvement would make the framework even more applicable and popularize it within the research field. This sets the frame for our future research.

5   Conclusion and Outlook on Future Research

This study was motivated by the need to improve the understanding of the phenomenon of FinTech, hyped by the media but quite ambiguous in IS research. Although, the first step forward establishing a common understanding has been made by the authors of [1], who analyzed FinTech seen from the perspective of the popular media and created the conceptual framework of its understanding, the question of its practical validity still remained open. Therefore, this study follows descriptive approach and addresses the topic of FinTech from different perspective with the purpose to discuss the evidences of FinTech transformation in five Swiss FinTech companies. These FinTech companies operate in several areas: crowdlending, digital payments, insurance, personal finance management and blockchain.

This work contributes to existing research and practice in several ways. It presents and discusses the evidences of FinTech transformation on the example five studied Swiss companies and, therefore, tests the conceptual framework, used as the theoretical background of the study. Furthermore, it extends the existing literature on FinTech and, therefore, contributes to financial and digital innovation literature.

Several limitations and opportunities for future research should be admitted. We will present limitations first, which go along with ideas for future research.

First, only a small number of Swiss FinTech companies was studied. Observing a larger variety of FinTech companies, operating in different areas, would be advantageous, as it could be used to make the framework a powerful tool to identify and cluster FinTech companies and innovations. The companies can be studied with the focus on different aspects, e.g. organizational structure and culture, business model, market orientation, willingness and readiness to cooperate, etc. Furthermore, by increasing the number of the companies, included in the study, one can examine the tendencies in FinTech over time.

Second, we should also admit that the study was conducted in order to examine Swiss FinTech arena, however, the results may look differently in other European countries and the ones in the Asian part of the world. The reason for that can be different working culture, regulatory system, economic conditions and a special place, which Switzerland takes on the world financial arena. Therefore, we encourage researchers, who are interested in the topic, to extend this research with a similar one, but addressing other important FinTech locations, and therefore providing a richer description of FinTech transformations happening. This will allow for more generalizability of the results to FinTech as an international phenomenon.

Third, in this research we have collected the information on the companies which is available in open access. However, we would recommend to advance this research by interviewing FinTech practitioners and presenting the opinions from "the wild". This could bring new opportunities and unveil unknown problems, important for FinTech insiders but invisible for the press, industrial observers and researchers.

Having these opportunities in mind, we are optimistic about future research directions and would like to encourage IS researchers to contribute more to the scientific literature on FinTech, which is currently on its rise but rather still underrepresented in research.

References

1. Zavolokina, L., Dolata, M., Schwabe, G.: FinTech - what's in a name? In: Thirty Seventh International Conference on Information Systems (2016)
2. Danneels, E.: Disruptive technology reconsidered: a critique and research agenda. J. Prod. Innov. Manag. **21**, 246–258 (2004)
3. Koellinger, P.: The relationship between technology, innovation, and firm performance— empirical evidence from e-business in Europe. Res. Policy **37**, 1317–1328 (2008)
4. CreditGate24 - Die innovative Direct Lending Plattform. https://www.creditgate24.com/
5. Creditgate24 100 Kreditprojekte und kein einziger Ausfall. Das grosse Interview mit der Schweizer Peer to Peer Lending Plattform|Der Finanzprodukt Blog. http://www.finanzprodukt.ch/finance-2-0/creditgate24-christoph-mueller-100-kreditprojekte-und-kein-einziger-ausfall-das-grosse-interview-mit-der-schweizer-peer-to-peer-lending-plattform/
6. Hypo Lenzburg steigt mit CreditGate24 ins Privatkreditgeschäft ein|cash. https://www.cash.ch/news/boersenticker-firmen/hypo-lenzburg-steigt-mit-creditgate24-ins-privatkreditgeschaft-ein-492658
7. Europe's Top 10 P 2P Lending Platforms (2016). http://fintechnews.ch/p2plending/europes-top-11-peer-to-peer-lending-platforms/4960/
8. Neue Anbieter locken mit Krediten zu Tiefstzinsen (2015). http://www.handelszeitung.ch/unternehmen/neue-anbieter-locken-mit-krediten-zu-tiefstzinsen-749313
9. Overview. https://www.twint.ch/en/about-us/overview/
10. Häufige Fragen – TWINT. https://www.twint.ch/support/faq/

11. Top 30 FinTech Startups in Switzerland. http://fintechnews.ch/top-30-fintech-startups-in-switzerland-2/
12. Gemeinsame Mobile-Payment-Lösung für die Schweiz beschlossen (2016). https://www.twint.ch/gemeinsame-mobile-payment-loesung/
13. Konkurrenz zu Twint: Postfinance rüstet App auf (2016). http://www.handelszeitung.ch/unternehmen/konkurrenz-zu-twint-postfinance-ruestet-app-auf-988450
14. Divé, M.: Über uns. http://www.knip.ch/ueber-uns/
15. Knip in Top 100 der besten Fintech-Unternehmen. https://www.knip.ch/fileadmin/user_upload/2015-12-16_Fintech100_CH.pdf
16. Knip|crunchbase. https://www.crunchbase.com/organization/knip#/entity
17. Fintech-Startup Knip erhält 14 Millionen Euro. https://www.knip.ch/fileadmin/user_upload/2015-10-26_Finanzierungsrunde_CH.pdf
18. Divé, M.: FAQ. http://www.knip.ch/faq/
19. Fintech-Star Knip bringt Versicherer gegen sich auf (2016). http://www.handelszeitung.ch/digitalisierung/fintech-star-knip-bringt-versicherer-gegen-sich-auf-1058353
20. Infografik zu InsurTech Finanzierungen inkl. 2 Schweizer Firmen (2016). http://fintechnews.ch/insurtech/infografik-zu-insurtech-finanzierungen-inkl-2-schweizer-firmen/2624/
21. Leimbach, A.: E-Banking der Zukunft. https://www.contovista.com/static/docs/Handelszeitung.pdf
22. Contovista lanciert Personal Finance Management Lösung - Erfolgreiche Kooperation zwischen Contovista, Zürcher Kantonalbank und Viseca. https://www.contovista.com/docs/Medienmitteilung.pdf
23. Contovista - Personal Finance Management for Banks. https://www.contovista.com/index/team
24. Aduno Group relies on Swiss start-up when it comes to digitisation. https://www.contovista.com/static/docs/201603XY_Media_release_Contovista_eng.pdf
25. Das sind die besten 100 Startups der Schweiz. http://www.limmattalerzeitung.ch/limmattal/region-limmattal/das-sind-die-besten-100-startups-der-schweiz-130569282
26. Ethereum Project. https://www.ethereum.org/
27. Contracts — Ethereum Homestead 0.1 documentation. http://ethdocs.org/en/latest/contracts-and-transactions/contracts.html#what-is-a-contract
28. What is Ether. https://www.ethereum.org/ether
29. About the Ethereum Foundation. https://www.ethereum.org/foundation
30. Ether – der bessere Bitcoin (2016). http://www.handelszeitung.ch/invest/ether-der-bessere-bitcoin-1014121
31. Wang, B.: Star Citizen Crowdfunding has raised Seven times more than biggest pre-2014 crowdfunding. http://www.nextbigfuture.com/2015/02/star-citizen-crowdfunding-has-raised.html
32. Sander, M.: Blockchain der Schweizer Ethereum-Stiftung: Ein Sündenfall als Antwort auf einen Hacker (2016). http://www.nzz.ch/wirtschaft/blockchain-der-schweizer-ethereum-stiftung-ein-suendenfall-als-antwort-auf-einen-hacker-ld.110502
33. Tatar, J.: Is Ethereum Becoming the New Platform for Startups? https://www.thebalance.com/is-ethereum-becoming-the-new-platform-for-startups-4018986

Reading Between the Lines: The Effect of Language Sentiment on Economic Indicators

Florian Förschler[1]([⊠]) and Simon Alfano[2]

[1] University of Tübingen, Tübingen, Germany
florian.foerschler@gmail.com
[2] University of Freiburg, Freiburg, Germany
simon.alfano@is.uni-freiburg.de

Abstract. Given the ever-increasing volume of information in financial markets, investors must rely on aggregated secondary data sources. Such data sources include indices such as the Ifo Business Climate Index in Germany or the Purchasing Manager Index (PMI) in the United States. However, such indices typically require one to interview experts and are thus cost-intensive and only published with a certain time lag. In contrast, we suggest evaluating the role of sentiment encoded in the mandatory, stock-relevant disclosures of stock-listed companies on various economic indicators. Such sentiment analysis builds on primary information, which covers a large share of the economy, comes at little cost and can reflect new information instantaneously. Our results suggest that such a sentiment analysis explains moves in stock indices and macroeconomic factors, namely the new order flow and unemployment rate.

Keywords: Economic indicators · Financial news · Sentiment analysis · Sentiment index

1 Introduction

The accelerating global expansion of the Internet over the last 20 years has transformed our economic reality. Increasing digitization has not only been reflected in the emergence of IT giants as eBay, Google or Amazon, but has also had a powerful impact on trading on electronic stock exchanges such as Xetra or Nasdaq [21].

The information revolution entails a lot of advantages for the existing economy, but challenges are becoming more obvious, too. Large amounts of data have become available, leading to an incredible growth in information, relevant for decision-making in financial markets. Still, the key challenge is to discern the pivotal pieces of information from the many irrelevancies in the abundant data pool [20].

Sentiment analysis is one approach to structure and quantify qualitative, textual pieces of information within this increasing flow of information. Sentiment refers to the optimism or pessimism embedded in language [34]. Sentiment analysis is a method of quantifying the extent of optimism or pessimism as conveyed by textual information such as news releases or financial disclosures [23].

© Springer International Publishing AG 2017
S. Feuerriegel and D. Neumann (Eds.): FinanceCom 2016, LNBIP 276, pp. 89–104, 2017.
DOI: 10.1007/978-3-319-52764-2_7

Text mining programs allow one to convert relevant data into information signals and utilize them for further analysis. The most important advantage of the digitization of the economy is the increasing efficiency of markets, which crucially depends on the availability of related data. Both producers and consumers also benefit from the widespread proliferation of information. Thanks to online commentaries and reviews, consumers can easily find out how other users evaluate a product. Producers profit from the ability of forecasting demand, changes in the economic environment and exchange rates more precisely.

The everyday business of stock exchanges has seen radical changes due to digitization, as well. Floor trading has all but disappeared as more trades are effected on electronic platforms. Now analysts and investors are subject to an increasing time pressure while analyzing myriads of isolated pieces of information and acting accordingly. Speed and quality of analysis are essential to the success of market participants. Liebmann et al. [21] demonstrate that analysts and investors vary in how long they need to digest new information. Under such circumstances, it becomes imperative to process incoming data as quickly and accurately as possible. The purpose of sentiment analysis is to scrutinize the latest ad hoc company news in a matter of mere seconds by turning a textual message into a numerical digit, the sentiment value.

In this context, this paper aims to study how the sentiment encoded in the mandatory, stock-relevant ad hoc filings of stock-listed companies conveys relevant information about the state of the economy. We evaluate how this sentiment measure can explain future movements in stock indices, the scale of industrial orders and the unemployment rate. Our findings suggest that sentiment can be a relevant indicator, but that its effect on economic indicators may shift during or after financial crises.

This paper aims at contributing to the understanding of how sentiment effects economic indicators and is embedded into an overall research project, which aims to develop a sentiment index. A sentiment index is advantageous compared to other indices, like the Ifo index or the American Purchasing Manager Index (PMI), because it is less cost-intensive to collect sentiment data and the method reflects new information instantaneously.

This paper's objective is to determine whether sentiment values are capable of explaining the performance of the CDAX and changes in macroeconomic variables. In Sect. 2, we review previous literature regarding sentiment analysis and indices. We introduce our research methodology in Sects. 3 and 4. In Sect. 5, we present our results and discuss their implications in Sect. 6, before concluding in Sect. 7.

2   Theoretical Background

2.1   Sentiment Analysis

As Arthur [1] has pointed out, judgment regarding current and future economic situations is highly influenced by subjective beliefs. Each individual forms expectations and hypotheses about the surrounding environment, which in turn provide the basis for economic decision-making. Moves on stock exchanges, therefore, result from market participants' perceptions of the future. Soroka [33] studies the impact of "good" and "bad" economic news and arrives at the conclusion that the reaction to bad news is considerably stronger.

Tetlock et al. [35] show that the influence of media on financial markets is substantial. Elevated levels of pessimism in the news exert downward pressure on exchanges. Nevertheless, a price correction, induced by negative emotions, is relatively quickly recovered as the market once again focuses on fundamentals. Tetlock supports the noise trader theory, which is described in greater detail below. Broadly, it postulates that market actors do not trade exclusively on fundamental data, but there is a portion of traders who decide between buying and selling based on the news and market sentiment.

Contemporary economic literature shows that news influences individual expectations which, in turn, propel the stock indices. For that reason, the research field in question is of importance when it comes to studying fluctuations in stocks prices. The process of surveying, analyzing and evaluating stock exchange sentiment is called sentiment analysis. This branch of research aspires to transform the news into numerical values using text and data mining approaches [21].

Ad hoc publications usually contain information about topics such as dividends, profit warnings, management changes or acquisition and divestiture activities. To duly evaluate publicly listed enterprises, analysts study all available information. Every public company publishes consolidated financial statements yearly, business reports quarterly, and many other releases about management, M&A and business strategy on an event-driven basis. One of the greatest challenges in terms of sentiment analysis, therefore, is to evaluate this variety of textual data [20].

Two factors that have promoted growth in this field of research are a tremendous increase in readily available electronic information due to the expansion of the Internet and advances in textual analysis techniques, which have made it possible to effectively study large bodies of textual information [20]. The intricacies of text analysis are a direct consequence of the polyvalent meanings of words when surrounded by other words. If a publication's title bears the name of a serious disease, it is weighted negatively. Yet, if the same disease is contextually related to a new drug developed by a company, the news item must be evaluated positively [22]. The value and relevance of the real-time analysis of data is evident and will doubtless reshape research methods in all branches of science [36]. In economics, sentiment analysis could soon contribute significantly to our understanding of economic developments.

2.2 Economic Indicators

Companies make investment decisions on the basis of their estimates of the future economic situation. If they anticipate an upswing, investments are likely to increase more – *ceteris paribus* – than if they predict a recession in which revenues would be lower. In every meeting held to discuss interest rates, central bankers should take into consideration how different macroeconomic aggregates will behave [2]. As realizations and expectations of inflation involve a larger number of determinants than merely GDP and unemployment – e.g. capacity utilization rate, home and asset prices, etc. – it is difficult to recognize which variables will end up in the central bank's model [6].

As Hayek states, markets are the most efficient ways of aggregating dispersed information [15]. If all necessary information is fully reflected in the prices, a market is considered efficient [7–9]. Friedman further popularized this concept in financial research, formalizing it in the efficient market hypothesis (EMH) [8,11]. But regarding the predictability of stock prices, there has been no unanimous point of view since the formulation of the EMH.

Logically, the EMH states that stocks are always fairly priced, and that it is impossible for market participants to score better returns than those of a benchmark index. With the throw of a dart, a chimpanzee would fare equally well (or poorly) as all educated stock market experts in naming a number of better-than-benchmark portfolios [24]. Malkiel demonstrates that the top 10 mutual funds that had beaten the returns of the S&P 500 in the 1960s, some by as much as double, lagged behind the same index in the 1970s [25]. Therefore, the current performance of the best mutual funds does not provide any clue about their future returns.

In "Random Walk Theory", Malkiel argues that stock prices incorporate every piece of information available during the day [25]. The EMH presumes perfect competition and complete information, an assumption which is often criticized. Likewise, there is no consensus about how far ahead macroeconomic developments can be forecasted. Elliot and Timmermann [6] conclude that time series models are unstable with the lapse of time, and argue that economic forecast models should rather be perceived as approximations.

Economic forecasts can be based on surveys (Ifo index, ZEW economic outlook) or on real-time data (sentiment analysis). The latter method utilizes a different approach as contrasted with the two most followed indicators in Germany. The boom in the technical advancement of information systems has considerably enriched the palette of approaches accessible to analysts in their research [6]. Due to a steady supply of real-time data, the accuracy of models can be tested more rapidly and efficiently. Thus, their performance can be proved or refuted in a timely fashion. With econometric models, it is worth noting that correlations are susceptible to change over a study period as a result of modifications in the framework (regulations, institutions, technology, etc.) or relevant occurrences (terror attacks, natural disasters, etc.). Hence, this type of model should allow for instability in its parameters. In general, there doesn't seem to be a single forecast model that produces reliable projections in practice [6].

Despite the constraints of the efficient market hypothesis, analysis of sentiment factors still offers a benefit that should not be underestimated. The efficiency of markets is crucially dependent on the speed of information processing. Data mining and text mining with regard to sentiment values make it possible to evaluate a news item almost at the moment of its release. In contrast, analysts and investors need considerably more time to grasp new information, evaluate it accordingly and effect a trade in the market [21]. Among other advantages, such a time edge makes sentiment analysis an approach worth pursuing.

3   Research Hypotheses

The sentiment of financial news is a potential measure of the optimism or pessimism of economic actors. Thus, it could serve as a leading indicator for other economic variables, such as stock indices, but also for macroeconomic indicators such as GDP or the unemployment rate. In this paper, we want to study the influence of sentiment on different economic measures in order to educate the selection of relevant economic indicators for a sentiment index, which we develop in a parallel paper. In this context, we develop the following hypotheses.

Since the dataset on market sentiment values consists of ad hoc news publications, there are reasonable grounds to assume that these news items exhibit a positive correlation with the CDAX, and that the respective sentiment values are capable of explaining the stock market performance. This hypothesis stems from the fact that the valuation of stocks is always influenced by market sentiment and expectations [30]. Siering followed a similar approach by testing whether positive news items have a greater impact on stock movements than negative ones [32].

Hypothesis 1 (H1): The constructed sentiment index is positively correlated with the performance of the CDAX.

Similar to the Ifo index and the ZEW economic outlook indicator, it is probable that monthly sentiment values demonstrate a relationship with economic development [18]. We additionally expect sentiment to explain macroeconomic indicators as our sentiment variable reflects the sentiment of all stock-relevant announcements from companies representing the leading German stock market, the CDAX. Therefore, this corporate news flow represents a major part of German business and is thus a good indicator for the pessimism or optimism prevalent among German companies. Our analysis will determine whether our sentiment variable is capable of explaining changes in macroeconomic variables (incoming orders, and unemployment level) and thus serve as a forecast indicator for fluctuations in economic activity.

Hypothesis 2 (H2): News sentiment correlates positively with the macroeconomic variable "incoming orders".

Hypothesis 3 (H3): News sentiment correlates negatively with the macroeconomic variable "unemployment rate".

4   Methodology

4.1   News Corpus

This section explains the choice of German ad hoc announcements in English as our underlying news corpus. To this end, we briefly recapitulate the corresponding publication regulations: to prevent insider trading and assure the equal availability of novel, stock-price-relevant information to all market participants, stock-listed firms are subject to disclosure regulations. Disclosure regulations must meet specific criteria regarding the content. However, the criteria do not regulate which words may be used. Thus, the choice of words (as the source of the conveyed sentiment) is left to the discretion of the news originator.

In Germany, the legal obligations ensure that companies disclose stock-price-relevant company information in the form of so-called ad hoc announcements [13,14,28]. The publications must include financial results, changes of top management, M&A transactions, major orders, dividends, and litigation outcomes. These regulated ad hoc disclosures are usually published in German and, more importantly for our analysis, in English. Each disclosure contains approximately 10–20 lines of free text. In contrast to the SEC-regulated publications in the US, companies in Germany have to file any new information immediately. This fact makes such disclosures highly relevant to stock market participants since the information they contain is definitively novel in nature [28]. In addition, ad hoc announcements need to be authorized by executives of the releasing company and the ad hoc filing is quality-checked by the Federal Financial Supervisory Authority[1]. Several publications have assessed their importance to stock market reactions and suggested a direct relationship with stock market returns [28].

Our news corpus consists of German regulated ad hoc announcements from between January 2004 and June 2011.[2] As a requirement, each announcement must be written in English. In addition, we remove so-called penny stocks with a value below 5 Euro, since these react less systematically to financial disclosures. In total, our final corpus consists of 14,427 ad hoc announcements.

4.2 News Sentiment Analysis

Sentiment analysis refers to analytical methods that measure the positivity or negativity of the content of textual data sources. In this way, sentiment analysis can shed light on how human agents process and respond to the textual content of news.

Before investigating the differential information processing of news sentiment, we need to pre-process our news corpus according to the following steps:

1. *Tokenization:* first, tokenization splits running text into single words, named tokens [10,26].
2. *Negation inversion:* We then account for negations using a rule-based approach to detect negation scopes and invert the meaning accordingly [4,31].
3. *Stop word removal:* in a next step, we remove so-called stop words, which are words without relevance, such as articles and pronouns [19].
4. *Stemming:* finally, we perform stemming in order to truncate all inflected words to their stem using the Porter stemming algorithm.

---

[1] Bundesanstalt für Finanzdienstleistungsaufsicht (BaFin).

[2] Kindly provided by Deutsche Gesellschaft für Ad-Hoc-Publizität (DGAP).

After completing the pre-processing, we can study the influence of news senti-ment on financial markets. For this purpose we choose the Net-Optimism met-ric [5] combined with *Henry's Finance-Specific Dictionary* [16], since this is a common sentiment approach that leads to a robust relationship [23]. The Net-Optimism metric $S(A)$ is given by the difference between the number of positive $W_{pos}(A)$ and negative $W_{neg}(A)$ words divided by the total number of words $W_{tot}(A)$ of an announcement A. Let us introduce the variables denoting news sentiment formally by

$$S(A) = \frac{W_{pos}(A) - W_{neg}(A)}{W_{tot}(A)} \in [-1; +1].$$

In addition, we aggregate the sentiment on a monthly basis across all ad hoc announcements within a given month.

4.3 Methodological Approach

To identify possible correlations between the sentiment values and CDAX perfor-mance, we initially compare data on a firm-to-firm basis, aggregate it afterwards on daily and monthly scales, and then test it for interconnections. For the pur-pose of demonstrating whether sentiment values are able to explain macroeco-nomic variables, a considerable amount of external data (new industrial orders, unemployment rate) is processed as well.

The first hypothesis is primarily validated through correlation tests and the OLS. The *Durbin-Watson test* and *Breusch-Pagan statistics* are applied to check for autocorrelation and heteroscedasticity. A regression corrected for the *Newey-West test* will produce more reliable results. Furthermore, by adding monthly lags, we estimate the extent to which sentiment values could be an early indicator for shifts in the dependent variables. The sentiment values, augmented by time lags and then analyzed, is treated as a control variable for the CDAX.

Hypotheses H2 and H3 are evaluated by Newey-West-corrected OLS regres-sions. The GDP, being the most important metric, cannot be a proper dependent variable because it is calculated quarterly. As an approximation of developments in GDP we use incoming orders. Another important macroeconomic indicator is the unemployment rate in Germany. These variables are examined for corre-lations with the sentiment index on a monthly basis. The dataset analyzed in this paper comprises a survey of ad hoc CDAX notifications from between 2004 and mid 2011. The CDAX, in contrast to the German DAX index, includes not only the 30 largest "blue chip companies", but over 400 small and medium-sized businesses, which makes it a better representative of Germany's economy. Con-stituents of the CDAX are all domestic stocks traded on the Frankfurt Stock Exchange, with foreign stocks falling outside the index range. Since the avail-able data derives from German companies only, hypotheses H2 and H3 are tested only in relation to German macroeconomic developments.

5   Results

5.1   Descriptive Statistics and Correlation Analysis

The corresponding descriptive statistics and variables used in the regression models are provided in Table 1. Our approach is to first test the correlation between sentiment values and CDAX data on an individual level.

Table 1. Descriptive statistics of time series (January 2004–June 2011).

Variable	Description	N	Mean	Std. dev.	Min	Max
$S(t)$	Sentiments	14,427	0.0001	0.0001	−0.001	0.004
$C(t)$	CDAX- Index	14,427	335.583	71.502	202.260	490.060
$\alpha(t)$	Alpha	14,427	0.080	1.211	−18.048	19.418
$AR(t)$	Abn. Ret	14,427	0.749	9.115	−80.022	311.566
$CAR(t)$	Cum. Abn. Ret	14,386	0.189	10.122	−132.841	190.531
$PB(t)$	P/B ratio	13,989	2.263	10.927	−385.660	281.910

Table 2 displays correlations between news sentiment and the CDAX. The arithmetic mean is the aggregation method. The result indicates a weakening of the sentiment effect upon aggregation. Sentiment values are highly correlated with the CDAX on a daily basis (p-value < 0.001) and still correlated at a statistically significant level on a monthly basis (p-value < 0.05). Since the data for the macroeconomic indicators evaluated under hypotheses 2 and 3 is only available on a monthly basis, we refer in the following sections to monthly data. As the daily aggregated sentiment values of the ad hoc announcements of companies listed on the CDAX have a larger correlation coefficient and are statistically more significant (p-value smaller than 0.01) than monthly aggregated sentiment values (p-value smaller than 0.05), this suggests the stronger statistical inference of a sentiment index on a daily basis. We build on this finding in a parallel paper, which focuses on the implementation of a daily sentiment index [17].

Table 2. Correlations between news sentiment and CDAX.

	Individual correlation	Daily correlation	Monthly correlation
News sentiment	0.068***	0.079***	0.235*
	(8.191)	(3.570)	(2.270)

Correlation Coeff., t- Stat.   Sign.: ***0.001, **0.01, *0.05

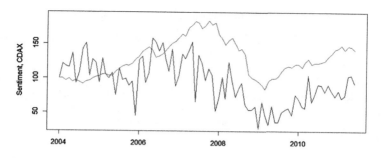

Fig. 1. Normalized sentiment and CDAX on a monthly basis (Jan. 2004 = 100).

5.2   The Effect of Sentiment on the CDAX

The following analysis tests whether the sentiment values are able to explain movements of financial markets (e.g. the CDAX).

Elliot concludes that it is necessary to analyze not only the whole time frame, but also to consider relations between variables within specific periods if there are macroeconomic shocks or major changes in institutional and legal frameworks [6]. As the financial crisis of 2007/08 was a significant macroeconomic shock, our analysis is subdivided into two periods: 2004 – Dec. 2007 and Dec. 2007 – 2011. This division is based on the fact that December 2007 represents the official beginning of the financial crisis according to the National Bureau of Economic Research [29] and marked a considerable change in the macroeconomic environment.

Sentiment values range within a certain interval, too, and for purposes of convenience we normalize the first monthly value in 2004 at 100. Figure 1 is a graphic illustration of the concept, where sentiment values are always absolute numbers (black line), not percentage changes. For this reason, we will compare the sentiment values to the CDAX itself (blue line), not its oscillations in terms of percentage. Below is the OLS regression model with which the defined hypothesis will be tested:

$$C(t) = \beta_0 + \beta_1 S(t) + \beta_2 \alpha(t) + \beta_3 AR(t) + \beta_4 CAR(t) + \beta_5 PB(t) \qquad (1)$$

where $C(t)$ is referred to as the CDAX. $S(t)$ reflects the sentiment variable, our independent variable of interest. In addition, the model includes several control variables: The market alpha $\alpha(t)$, abnormal returns $AR(t)$, the cumulative abnormal return $CAR(t)$ and the price-to-book-ratio $PB(t)$. Results of autocorrelation and heteroscedasticity tests are listed in Table 3 and suggest that both effects are apparent in the dataset and that the initial regression should be corrected. Table 4 summarizes the products of the regression augmented by the Newey-West test to avoid autocorrelation and heteroscedasticity.

The analysis shows a statistically significant coefficient of sentiment for the period after the financial crisis (p-value < 0.001). If we take all six sentiment

Table 3. Results of a test on autocorrelation and heteroscedasticity of certain variables.

	CDAX	New orders	Unemployment rate
Durbin-Watson test	0.273***	0.683***	0.650***
Breusch-Pagan test	16.572*	14.422*	6.700

Correlation Coeff., t- Stat.　　　　　　　　Sign.: ***0.001, **0.01, *0.05

Table 4. Results of a Newey-West corrected regression of sentiment index and CDAX for several time periods.

	(1) (2004–2007)	(2) (2004–2007)	(3) (2007–2011)	(4) (2007–2011)	(5) (2004–2011)	(6) (2004–2011)
(Intercept)	0.000***	0.000***	0.000***	0.000***	0.000***	0.000***
	(4.086)	(3.781)	(8.055)	(1.950)	(8.372)	(3.955)
S(t)	−0.054	−0.162	0.466***	0.243**	0.232	0.003
	(−0.315)	(−0.997)	(7.000)	(2.945)	(1.158)	(0.018)
S(t − 1)		−0.025		0.103		0.053
		(−0.0323)		(1.204)		(0.460)
S(t − 2)		0.063		0.248**		0.152
		(0.571)		(3.495)		(1.620)
S(t − 3)		0.059		0.195*		0.066
		(0.726)		(2.436)		(0.841)
S(t − 4)		−0.091		0.207**		0.016
		(−1.150)		(3.360)		(0.169)
S(t − 5)		0.013		0.127		0.073
		(0.138)		(1.963)		(0.658)
S(t − 6)		0.226		−0.045		0.148
		(1.530)		(−0.056)		(0.895)
$\alpha(t)$	−0.242	−0.295	−0.147	0.176*	−0.155	−0.076
	(−0.483)	(−1.483)	(−1.323)	(2.180)	(−0.078)	(−0.523)
AR(t)	−0.134	−0.183	0.122	0.116*	−0.132	−0.087
	(−1.208)	(−1.989)	(1.846)	(2.520)	(−1.542)	(−0.830)
CAR(t)	−0.152	0.162	−0.099	−0.242***	−0.003	−0.107
	(0.558)	(0.882)	(−1.248)	(−5.525)	(−0.019)	(−1.110)
PB(t)	−0.233	−0.345**	0.456***	−0.242*	−0.010	−0.107
	(−1.397)	(−2.776)	(4.353)	(2.598)	(−0.631)	(−1.136)
R^2	0.110	0.211	0.593	0.807	0.111	0.216
Adj. R^2	0.005	0.000	0.538	0.739	0.058	0.096

Correlation Coeff., t- Stat.　　　　　　　　Sign.: ***0.001, **0.01, *0.05

lags into account, we also get significant results for Lag 2 through 4, which corroborates the previous lag analysis presented in Table 4. The sentiment index is able to explain CDAX movements over the next 4 months. **Hypothesis (H1)** therefore cannot be rejected for the (post-) crisis observation period, whereas it has to be rejected for the previous time frame and also for the whole period of 2004–2011, since the results are not statistically significant within these time spans. An explanation for the results after the financial crisis may be that financial news was keenly watched due to the massive impact of collapsing markets and high volatility in the aftermath [3].

5.3 Sentiment Index and Macroeconomic Indicators

The Ifo index and ZEW economic outlook are said to be able to forecast certain economic developments. The prediction of the near future is defined as nowcasting. In this section, we test whether the news sentiment index possesses a similar nowcasting potential as the Ifo index and ZEW outlook [12,27] by studying the effect of news sentiment on certain economic indicators.

Entrepreneurs usually make investment decisions based on forecasts of the future economic situation, whereas the current environment feeds into such predictions. If the economy is in a recession at present, businessmen tend towards pessimism in the short term and refrain from embarking on new projects. We also suppose the inverse: If the news and market sentiment are optimistic, businesses should be observed to be eager to invest. In macroeconomics, the most influential metric is, of course, the GDP of a country. However, this metric is estimated only quarterly, and thus we have to recourse to another economic indicator – incoming orders – which we will use as an approximation of German economic development. Tests are carried out on a monthly basis.

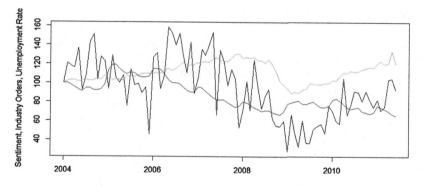

Fig. 2. Normalized sentiment, incoming orders in the industrial sector (green) and unemployment rate (blue). (Color figure online)

In principle, our analysis in this part follows the approach of Huefner [18], whereby the incoming industrial orders, IO(t), function as the dependent variable. Furthermore, the regression below will also check whether the sentiment index can explain fluctuations in the unemployment rate, UR(t). With independent variables remaining the same, the updated model may be seen below:

$$IO(t)/UR(t) = \beta_0 + \beta_1 S(t) + \beta_2 \alpha(t) + \beta_3 AR(t) + \beta_4 CAR(t) + \beta_5 PB(t) \quad (2)$$

Building on Fig. 2, one may suppose a positive correlation between the two datasets. The sentiment values will be colored in black and macroeconomic variables in blue and green in all figures that follow. For the sake of convenience, the datasets have been normalized, and graphs start at 100 in 2004. It is logical to assume that our regression should be more successful at explaining incoming orders than, for example, industrial production. Due to purchasing and planning procedures, industrial production is always protracted. The outcome in Table 5 shows that we obtain slightly significant positive values for the sentiment index in column 5 (p-value < 0.05).

By adding control variables three to five, the sentiment index shows a significant effect on incoming industrial orders. Therefore, we cannot reject **Hypothesis (H2)**.

Finally, we set the unemployment rate as the dependent variable. In this case, the coefficient for sentiment in the first column is significantly positive (P-value < 0.001), and remains significant by adding further control variables (Table 6). The adjusted R^2 accounts for 22.6% of explained variation in residuals.

The result is rather counter-intuitive, as we expect unemployment to decrease when we observe a positive sentiment. A possible explanation for this phenomenon is that since the financial crisis, news sentiment has been rather negative, but the German unemployment rate did not increase as in most other European countries. Unlike many other European countries, Germany saw its unemployment plunge in the aftermath of the financial crisis. The political measures in the wake of the crisis did not lead to massive unemployment but, on the contrary, to a further decrease in unemployment, which might explain the positive rather than negative correlation in this analysis.

The results of our analysis confirm that sentiment values can explain changes in the unemployment rate. However, we have to reject **Hypothesis (H3)**, since the coefficient of sentiment is positive, and not negative as hypothesized.

This result seems counterintuitive at first glance, but may be due to the special situation of the German labor market during the observation period: unemployment was relatively high at the beginning of our observation period, when the news sentiment was also relatively high, as companies had not yet been affected by the financial crisis. Later on, the unemployment rate was kept at lower levels during the financial crisis due to various labor market measures in Germany that contained the threat of rising unemployment during the crisis. At the same time, the sentiment was low during this period.

6   Discussion and Implications for Future Research

Our results demonstrate the poly-valence of the sentiment derived from the financial disclosures of stock-listed companies. This sentiment measure has a statistically significant effect on various economic indicators, including the German CDAX stock market, as well as other macroeconomic indicators such as the level of new orders and the unemployment rate. For indicators, which are available on a daily resolution, it is recommendable to evaluate the relationship

Table 5. Results of the Newey-West OLS regression for the relationship between the new industrial orders and the sentiment index.

	(1)	(2)	(3)	(4)	(5)
(Intercept)	0.000***	0.000***	0.000***	0.000***	0.000***
	(8.944)	(15.085)	(16.865)	(16.861)	(15.115)
S(t)	0.334	0.313	0.309*	0.318*	0.318*
	(1.426)	(1.905)	(2.0935)	(2.188)	(2.077)
$\alpha(t)$		−0.305***	−0.298**	−0.328*	−0.329*
		(−3.432)	(−3.349)	(−2.197)	(−2.164)
AR(t)			−0.100	−0.109	−0.108
			(−0.849)	(−0.866)	(−0.868)
CAR(t)				0.065	0.065
				(0.458)	(0.448)
PB(t)					0.002
					(0.015)
R^2	0.112	0.205	0.214	0.218	0.218
Adj. R^2	0.102	0.186	0.187	0.181	0.171

Correlation Coeff., t- Stat.   Sign.: ***0.001, **0.01, *0.05

Table 6. Results of the Newey-West OLS regression for the relationship between unemployment rate and the sentiment index.

	(1)	(2)	(3)	(4)	(5)
(Intercept)	0.000***	0.000***	0.000***	0.000***	0.000***
	(14.810)	(16.447)	(16.698)	(16.718)	(16.103)
S(t)	0.476***	0.499***	0.497***	0.483***	0.476***
	(5.447)	(6.250)	(6.348)	(6.584)	(5.819)
$\alpha(t)$		0.323***	0.327***	0.369***	0.359**
		(3.843)	(3.642)	(3.812)	(3.189)
AR(t)			−0.048	−0.035	−0.027
			(−0.761)	(−0.483)	(−0.347)
CAR(t)				−0.092	−0.091
				(−1.167)	(−1.099)
PB(t)					−0.055
					(0.694)
R^2	0.226	0.330	0.332	0.339	0.341
Adj. R^2	0.217	0.315	0.309	0.307	0.302

Correlation Coeff., t- Stat.   Sign.: ***0.001, **0.01, *0.05

on a daily basis, as the effect of sentiment on stock markets is more pronounced when looking at narrower time windows.

Thus, future research on this topic may evaluate shorter time windows, e.g. daily data, when constructing a sentiment index for the CDAX. In addition, in order to provide a useful tool for practitioners, researchers should also focus on developing an early-warning system, which translates the sentiment index into a useful sentiment prediction tool. Within the broader horizons of our overarching research project, we have set out to refine the properties of such a sentiment index for stock markets [17].

7   Conclusion

Sentiment analysis can provide valuable insights into relevant information signals for investors, in addition to fundamental information sources. In comparison to other indices, a sentiment index benefits from faster processing of available information. We have also devised a method of creating an index from data on market sentiment, calculating it as an arithmetic mean of daily and monthly sentiment values. Our analysis reveals interesting results regarding the usage of a monthly sentiment index.

First, monthly sentiment values are able to explain movements in the leading German stock index, the CDAX, at statistically significant levels during and after the financial crisis, but not before. This confirms prior research in demonstrating that the perception of sentiment may be subject to shifts. Second, the constructed sentiment metric remains a statistically significant predictor of the CDAX for up to four time lags. Third, sentiment is also able to explain changes of relevant macroeconomic variables, namely the volume of incoming orders. Thus, our sentiment values explain both stock market movements and macroeconomic developments.

References

1. Arthur, W.B.: Complexity in Economic and Financial Markets. Univeristy of Stanford, California (1995)
2. Blanchard, O., Amighini, A., Giavazzi, F.: Macroeconomics: A European Perspective, vol. 2 (2013)
3. Caporale, G.M., Spagnolo, F., Spagnolo, N.: Macro news and stock returns in the Euro area: a VAR-GARCH-in-mean analysis. Int. Rev. Finan. Anal. **45**, 180–188 (2016)
4. Dadvar, M., Hauff, C., de Jong, F.: Scope of negation detection in sentiment analysis. In: Proceedings of the Dutch-Belgian Information Retrieval Workshop (DIR 2011), Amsterdam, Netherlands, pp. 16–20 (2011)
5. Demers, E.A., Vega, C.: Soft information in earnings announcements: news or noise? SSRN Electron. J. (INSEAD Working Paper No. 2010/33/AC) (2010)
6. Elliot, G., Timmermann, A.: Economic forecasting. J. Econ. Lit. **46**, 3–56 (2008)
7. Fama, E.F.: The behavior of stock-market prices. J. Bus. **38**(1), 34–105 (1965)

8. Fama, E.F.: Efficient capital markets: a review of theory and empirical work. J. Finan. **25**(2), 383–417 (1970)
9. Fama, E.F.: Market efficiency, long-term returns, and behavioral finance. J. Finan. Econ. **49**(3), 283–306 (1998)
10. Feuerriegel, S., Neumann, D.: News or noise? How news drives commodity prices. In: Proceedings of the 34th International Conference on Information Systems (ICIS 2013). Association for Information Systems (2013)
11. Friedman, M.: Essays in Positive Economics. University of Chicago Press, Chicago (1953)
12. Giannone, D., Reichlin, L., Small, D.: Nowcasting: the real-time informational content of macroeconomic data. J. Monetary Econ. **55**, 665–676 (2008)
13. Groth, S.S., Muntermann, J.: An intraday market risk management approach based on textual analysis. Decis. Support Syst. **50**(4), 680–691 (2011)
14. Groth, S.S., Siering, M., Gomber, P.: How to enable automated trading engines to cope with news-related liquidity shocks? Extracting signals from unstructured data. Decis. Support Syst. **62**, 32–42 (2014)
15. Hayek, F.A.: The use of knowledge in society. Am. Econ. Rev. **35**(4), 519–530 (1945)
16. Henry, E.: Are investors influenced by how earnings press releases are written? J. Bus. Commun. **45**(4), 363–407 (2008)
17. Krinitz, J., Alfano, S., Neumann, D.: How the market detect its own mispricing - a news sentiment index to detect irrational exuberence. In: Proceedings of the 50th Hawaii International Conference on System Sciences (HICSS), Waikoloa, Big Island, 4–7 January 2017
18. Lahl, D., Huefner, F.: What determines the ZEW indicator? Centre for European Economic Research 03–48 (2003)
19. Lewis, D., Yang, Y., Rose, T., Li, F.: RCV1: a new benchmark collection for text categorization research. J. Mach. Learn. Res. **5**, 361–397 (2004)
20. Li, F.: Market analysis of corporate disclosures: a survey of the literature. J. Account. Lit. **29**, 143–165 (2010)
21. Liebmann, M., Hagenau, M., Neumann, D.: Information processing in electronic market: measuring subjective interpretation using sentiment analysis. In: Proceedings of the Thirty Third International Conference on Information Systems, Orlando (2012)
22. Loughran, T., McDonald, B.: When is a liability not a liablity? Textual analysis, dictionaires, and 10- ks. J. Finance **66**, 35–65 (2011)
23. Loughran, T., McDonald, B.: Textual analysis in accounting and finance: a survey (2016)
24. Malkiel, B.G.: The efficient market hypothesis and its critics. J. Econ. Perspect. **17**, 59–82 (2003)
25. Malkiel, B.G.: Reflections on the efficient market hypothesis: 30 years later. Finan. Rev. **40**, 1–9 (2005)
26. Manning, C.D., Schütze, H.: Foundations of Statistical Natural Language Processing. MIT Press, Cambridge (1999)
27. Marta Banbura, D.G., Modugno, M., Reichlin, L.: Now-casting and the real-time data flow. European Central Bank 1564, July 2013
28. Muntermann, J., Guettler, A.: Intraday stock price effects of ad hoc disclosures: the German case. J. Int. Finan. Markets Institutions Money **17**(1), 1–24 (2007)
29. National Bureau of Economic Research: US business cycle expansions and contractions (2016). http://www.nber.org/cycles.html

30. Porta, R.L.: Expectations and the cross- section of stock returns. J. Finance **51**, 1715–1742 (1996)
31. Pröllochs, N., Feuerriegel, S., Neumann, D.: Enhancing sentiment analysis of financial news by detecting negation scopes. In: Proceedings of the 48th Hawaii International Conference on System Sciences (HICSS). IEEE Computer Society (2015)
32. Siering, M.: Boom or ruin - does it make a difference? Using text mining and sentiment analysis to support intraday investment decisions. In: Proceedings of the 45th Hawaii International Conference on System Sciences (HICSS), pp. 1050–1059 (2012)
33. Soroka, S.N.: Good news and bad news: asymmetric responses to economic information. J. Polit.(2006)
34. Tetlock, P.C.: Giving content to investor sentiment: the role of media in the stock market. J. Finance **62**(3), 1139–1168 (2007)
35. Tetlock, P.C., Saar-Tsechansky, M., Macskassy, S.: More than words: quantifying language to measure firms' fundamentals. J. Finance **63**, 1437–1467 (2008)
36. Wu, X., Zhu, X., Wu, G.Q., Ding, W.: Data mining with big data. Department of Computer Science, University of Massachusetts

Cashless Society: When Will Merchants Stop Accepting Cash in Sweden - A Research Model

Niklas Arvidsson[1], Jonas Hedman[2(✉)], and Björn Segendorf[3]

[1] KTH Royal Institute of Technology, Stockholm, Sweden
niklas.arvidsson@indek.kth.se
[2] Copenhagen Business School, Frederiksberg, Denmark
jh.itm@cbs.dk
[3] Sveriges Riksbank, Stockholm, Sweden
bjorn.segendorf@riksbank.se

Abstract. Over the past decades, we have witnessed changes into how individual's pay. In particular, there has been a drop in the use of cash as payment instrument both in terms of value and frequency. Consequently, the amount of outstanding cash is shrinking. For instance, in Sweden the level of cash is around 1.5% of Gross Domestic Product. This might be a tipping point for when cash is of practical use. In the paper, we present a research model that explores when merchants will stop accepting cash.

Keywords: Cashless society · Merchants · Cash adoption

1 Introduction

Payments are essential in the exchange of money for goods and services between sellers and buyers. The most used payment instruments in point of sales locations are cash and payment cards. Over the past decades, we have witnessed changes into how individual's pay. In particular, there has been a drop in the use of cash as payment instrument both in terms of value and frequency. Payment cards, such as charge, credit, and debit, and more recently new payment instruments, such as mobile payments and e-money, are replacing cash. These changes occur more or less in all economies and across the globe, but are particular evident in the Nordic countries, where you also can find a lively debate on the cashless society [3, 4, 8]. For instance, in Sweden there is a cash rebellion "Kontantuppror", where lobby groups in particular representing the cash-in-transit service industry and older people, demand that banks accept cash again (Note that most Swedish bank branch offices are cashless).

In parallel, payments are receiving increased attention from academic communities and span several disciplines, including information systems [11, 20], consumer research [10, 23, 25], marketing [17, 26], economics [6, 22], sociology [16], management science [2, 3, 24], and banking and finance [12, 15]. This has resulted in a variety of topics in the study of payments, including what money is [29], cost-benefit analysis of cash and payment cards [7, 27], competition [9], social implications [18], choice and spending behavior [25, 27], payment framework [5], and adoption of mobile payments [1, 20, 31].

© Springer International Publishing AG 2017
S. Feuerriegel and D. Neumann (Eds.): FinanceCom 2016, LNBIP 276, pp. 105–113, 2017.
DOI: 10.1007/978-3-319-52764-2_8

Despite the above, one aspect of payment research which has been largely ignored is merchant acceptance of payments, i.e. why do merchants accept or don't accept specific payment instruments. There are some exceptions, including the adoption of mobile payments by merchants [21] and the study of merchants point of sales data [26]. One finding is that cash payments are much more expensive than card payments [6] and we witness a "...*movement toward greater use of electronic payment methods, though gradual, is uniform and unmistakable, both across countries and over time*" [14, p. 936]. Schreft [28, p. 5] puts forward critic on existing research "...*is backward looking. It tells us what payment instruments were chosen in the past may not be a good indicator of what will be chosen [accepted] in the future*". In the realm of an emerging cashless society, we are in particular interested in when merchants stop accepting cash.

We assume that merchants are economic rationale in their decision making, i.e. merchants will stop accepting cash when it becomes more expensive to manage cash acceptance than the marginal profit on cash sales. It is important to note that Swedish merchants are not legally bound[1] to accept cash as a mean of payment but can decide themselves which payment services to accept. However, we acknowledge the existence of other factors influencing this choice, including the risk of being robbed.

Our work has the potential of contributing to the understanding of merchant's role towards a cashless society by developing a research model that explains the when merchants will stop accepting cash at point of sales.

2   Background

The context of this study is Sweden, since it is among the countries in the world with the lowest value in banknotes and coins in circulation compared with gross domestic product.

2.1   Retail Payments in Sweden

One measure of cash use is the value of outstanding cash – bills and coins – compared with the gross domestic product (GDP), which varies between countries, as shown in Table 1. This measure provides an estimate of how dependent a payment system on cash, since many cash transactions are person to person (P2P) transactions and therefore not registered in any official statistics. The numbers for Sweden show a long-term downward trend when comparing outstanding cash to GDP. In 1950 this number was nearly 10 percent but the last ten years it has been below three percent, and 1,8 percent in 2015, as shown in Table 2. The most recent statistics show that the

---

[1] The Riksbank law states that cash is legal tender in Sweden and should therefore be accepted, but the freedom to enter contracts underpinning contractual and commercial law implies that a payer and a payee can enter an agreement that sets the Riksbank law aside. It should be noted that there are few court case rulings in this area and none in the highest court. This is not the case in Denmark or Norway, where central bank laws have superiority over contractual and commercial law.

Table 1. Outstanding cash in selected countries 2014, some selected countries

Countries/regions	Cash-in-circulation as share of GDP (M_0/GDP; %)
Malaysia	102,4
Chile	30,2
Bulgaria	12,2
Czech Republic	10,1
Euro-zone	9,7
Pakistan	8,8
USA[a]	7,1
UK	3,5
Denmark	3,0
Norway	2,0[b]
Sweden	2,0

Sources: European Central Bank (ECB), Norges Bank, Sveriges Riksbank and www.knoema.com. [a]The figures for the US is from the year 2010.
[b]This number is based on Norway's main-land GDP, i.e. excluding the off-shore oil sector.

Table 2. Value of banknotes and coins in circulation (annual averages; banks' holdings excluded)

	2006	2007	2008	2009	2010	2011	2012	2013	2014	2015
Value in billion SEK	96,5	97,0	96,7	96,5	95,5	90,7	86,8	84,4	78,2	74,9
Value as share of GDP	3,1	2,9	2,9	2,9	2,7	2,5	2,4	2,2	2,0	1,8

Source: The Swedish Financial Market 2015, The Riksbank (www.riksbank.se/en/) and Statistics Sweden.

number is around 1,5 percent in August, 2016. This long-term decline is, however, often the result of a process where GDP is increasing faster than the outstanding value of cash. A second, and quite extraordinary observation, is that the nominal outstanding value of cash has been declining since its peak in 2007. This is – to our knowledge – unique for Sweden and a strong indication of the transition towards a more or less cashless society in Sweden. According to data from the Riksbank, the decline is significant and fast also in 2016 where the nominal value of outstanding cash decreased over 12 percent in the period from January to July.[2]

The statistics on the usage of payment instruments in Sweden show that cash is rapidly declining. There several explanations for this. The first explanation is the long-term increase in the use of card payments in Sweden. Card payment schemas were launched in a greater scale during the 1990s and merchants as well as consumers have

---

[2] It should be mentioned that Sweden is currently replacing all banknotes and most coins with new ones. The direct short-term impact of this on the value of currency in circulation is ambiguous but it does not affect the strong long-term negative trend.

adopted card payments as the most important payment instrument in retail point of sales (POS) locations. The number of card transactions at POS and the value of these transaction have been growing steadily during the last ten years. In addition, the number of ATM withdrawals and value of such withdrawals have been declining over that last ten years (Table 3). Finally, survey undertaken by Sveriges Riksbank indicates that the use of cash at the point of sale in terms of volume of payments has fallen from close to 40 per cent to close to 20 per cent between 2010 and 2014. A recent, however not yet published, study by the Riksbank indicates that the number is down to around 15 per cent in 2016.

Table 3. Card transactions in payment terminals (POS) and ATM withdrawals

	2006	2007	2008	2009	2010	2011	2012	2013	2014	2015
ATMs										
No. of ATMs	2816	3085	3236	3319	3351	3566	3416	3237	3231	3285
Transactions (millions)	313	320	295	269	241	221	210	183	167	154
Value (SEK billion)	270	240	239	232	225	206	190	174	171	153
Payment card terminals										
Terminals (thousands)	184,6	187,3	194,8	217,8	203,1	209,6	198,4	195,8	197,0	183,8
Transactions (millions)	1000	1154	1358	1490	1646	1799	2048	2329	2423	2501
Value (SEK billion)	423	463	488	496	557	598	654	722	754	747

Source: The Swedish Financial Market 2015, The Riksbank (www.riksbank.se/en/).

A second explanation is the recent is the introduction of a mobile payment service that enables real-time clearing and settlement and therefore provide similar functionality as cash, i.e. the value in the transaction is transferred in real-time from the payer to the payee. Much in analogy with the passing of a bill from one hand to the other. This service – which is called Swish – was launched by banks in 2012 and has become very popular for person-to-person payments. The growth was high in both 2014 and 2015. By the end of August 2016, the service was used by 4,7 million Swedes and transactions worth 8,2 billion SEK were made during August 2016[3]. This service has, however, not yet been adopted at a large scale by merchants.

2.2 Merchants' Choice of Payment Services

Merchants in Sweden can decide to not accept cash but are at the same time not allowed to issue a surcharge to consumers related to which payment service that are used, which in essence means that merchants' choice of ideal payment service primarily is based on the direct costs and revenues related to each specific payment service. But research also shows that other factors play an important role as well.

---

[3] See www.getswish.se.

A study by Loke [19] of which factors that determine a merchant's decision to participate in a card scheme identified the following factors as important explanatory variables:

- Merchant's background (including age, number of personal credit cards held[4], and use of computers[5])
- Merchant's business characteristics (including business sector, total value of transactions per month, average value of transaction, profit margin, location of business)
- Effects of other players' decision via merchant's perception (including merchant's perception of customers' use of cards and competitors' participation in the card scheme)

The study (ibid) arrived at two main conclusions. The quantitative analysis related to the factors above showed that the statistically significant explanatory variables were: age of the merchant which had a negative relation to the probability of accepting card payments; number of cards held by merchant (positive relation); business sector (where surprisingly enough non-technical stores were more positive to accepting cards); total value of transactions (positive relation); merchants' perception of customers' use of cards (positive relation); and competitors' acceptance of card payments (positive relation). When discussing these results, the study concluded that the demand from customers was the most important factor while the merchants' wish to boost sales related to acceptance of card payments was the second most important factor.

Other studies [12, 13, 30] highlight the importance of different characteristics of merchants, characteristics of payers or consumers and a number of other factors. Regarding the characteristics and decision factors of merchants this includes: industry, location, margins or profitability, type of products sold, type of customers, price of payment services, amount of revenue connected to payment service, inter-operability of the service, as well as other factors such as, for instance, risks, employees' opinions and work environment issues. Regarding the characteristics and decision factors of payers this includes: socio-demographic characteristics (age, gender, education, income), transaction frequency and value, speed and ease of use of services, need for integrity, technology interest, trust in services and in the payment system in general, costs of payment and banking services, as well as other factors.

3 Research Model

Based on the review of retail payments in Sweden our research model is therefore rather straightforward. We ask ourselves if and under which circumstance merchants will stop accepting cash and instead accept only card payments, and base the analysis on business factors such as revenues and costs related to cash and revenues and costs related to card payments. Our model is based on the following features:

---

[4] A proxy for experience with cards.

[5] A proxy for familiarity with new technologies.

Let θ be a set of payment methods that can be used at the point of sale. For simplicity, assume that there is only two such methods a and b. One may think of a as cash and b as cards. Each merchant chose whether to accept cash (a = 1) and cards (b = 1). If he does not accept cash (cards) then a = 0 (b = 0).

Let the continuous function ri:θ → R₊ denote the revenue merchant i makes depending on his choice of payment methods. In particular, let $r_i(0,0) = 0$. Accepting no payment methods can be interpreted as a decision to exit the market. Here, for simplicity, we will initially assume the specific functional form $r_i(a, b) = r_i^a * a + r_i^b * b$, i.e. the revenue generated from card payments does not change if merchant also accept cash and vice versa.

Let π_i denote the profit of the merchant, cj the fixed cost and vj the variable cost for each merchant of accepting payment methods j = a, b. With variable cost, we mean the cost as a function of the size of the revenue generated by the payment method in question. We also assume that all merchants have access to the same technology for receiving payments. In Sweden, this can be motivated by nearly all points of sale have to use cash registers that are approved by the tax authority, i.e. the tax authority limits the merchants' choice to a narrow and clearly defined set of technologies.

Let $x \in (0,1)$ denote the revenue margin, i.e. the share of the revenue that exceeds the cost of the good sold. The merchants profit maximization problem is thus to maximize π_i over θ.

$$Max\ \pi_i(a, b) = a * r_i^a(x - v_a) + b * r_i^b(x - v_b) - c_a * a - c_b * b$$

In the general case this means that the merchant will accept cash if $r_i^a(x - v_a) - c_a \geq 0$ for b = 0 or 1 or both. In the following we will restrict ourselves to the case where merchants already accept cards (b = 1). The first reason for this is that it allows us to focus on the question of when does a merchant abandon cash and still remains on the market. The second reason is that in Sweden cards are nearly universally accepted.

To answer our research question, we have to estimate x, v_a and c_a which will allow us to find the critical revenue threshold

$$r^a* = c_a/(x - v_a)$$

which the merchant accepts cash and below which he does not.

3.1   Selected Survey Items and Points for Discussion

Data collection will be carried out by the three big trade associations, covering retailing and restaurants. The unit of data collection is a store (location) in 2016.

- What is the total turnover incl. VAT in 2016 of your store_____?
  Total value in cash_____? And share of total transactions (%)_____?
  Total value in card_____? And share of total transactions (%)_____?
- What is the shares of **transactions size** related to value (%) 1−19 SEK_____? 20−99 SEK_____? 100−499 SEK___? And 500-higher SEK_____?

- What is the distribution of your costumers in **age** in percent? Children (0–11 years old)____Youth (12–17 years old)____ Adults (18–65 years old)____Retired (over 65 years old)____
- How large is your **profit margin** in percent? (We will use industry averages)
- Do you **accept cash**? ____ (Yes/No)

Cost for Accepting Cash.

- What is the **average hourly labour cost** per hour per employee at your business (We will use industry averages)? ____
- Estimate your **total costs** in 2016 that are related to cash handling?____
- What was the **cost of cash** due to, for instance incorrect change, theft, forged cash, robberies, etc. that was not covered by insurance?____
- Try to estimate how much time per day the **employees devote to count cash**. Provide an estimate of a daily average____hours and____minutes per day?
- Estimate the **costs of cash storage** (depreciation of safety vaults, fees to baks and cash depots, etc.) ____SEK in storage fees 2016?

Substitution Effect.

- How much did your company **pay the bank/card company** on average on top of the fixed costs? Per transaction (on average) or fee + _____% of the amount

Background Questions.

- Do you have employees that sometimes work **alone** in a store? ____(Yes/No)
- Do you consider **stop accepting** cash? ____(Yes/No)
- Would your **sales decrease** if you stopped accepting cash?____(Yes/No)
- Would your **profits decrease** if you stopped accepting cash?___(Yes/No)
- Which year do you think you will **stop accepting** cash? ____
- Has your business suffered from **robberies** during the last five (or ten) years?____ (Yes/No)
- Do you know if any other store that is close to your stores suffered from **robberies** during the last five (or ten) years? _____(Yes/No)
- Is it common for companies in your industry to **pay under the table**?____(Yes/No)

References

1. Arvidsson, N.: Consumer attitudes on mobile payment services-results from a proof of concept test. Int. J. Bank Mark. 32(2), 150–170 (2014)
2. Arvidsson, N.: Det kontantlösa samhället: rapport från ett forskningsprojekt (2013)
3. Arvidsson, N.: Proceedings Third International Cashless Society Roundtable (ICSR), Stockholm (2013)
4. Carton, F., Hedman, J.: Proceedings: Second Internationael Cashless Society Roundtable (ICSR), Financial Services Innovation Centre (2013)
5. Carton, F., Hedman, J., Damsgaard, J., Tan, K.-T., McCarthy, J.: Framework for mobile payments integration. Electron. J. Inf. Syst. Eval. 15(1), 14–25 (2012)

6. Garcia-Swartz, D., Hahn, R., Layne-Farrar, A.: The move toward a cashless society: a closer look at payment instrument economics. Rev. Netw. Econ. **5**(2), 175–197 (2006)
7. Garcia-Swartz, D.D., Hahn, R.W., Layne-Farrar, A.: The economics of a cashless society: an analysis of the costs and benefits of payment instruments. AEI-Brookings Joint Center for Regulatory Studies (2004)
8. Hedman, J.: Proceedings First International Cashless Society Roundtable (ICSR), Copenhagen, Denmark, 18 & 19 April. Copenhagen Finance IT Region (2012)
9. Hedman, J., Henningsson, S.: The new normal: market cooperation in the mobile payments ecosystem. Electron. Commer. Res. Appl. **14**(5), 305–318 (2015)
10. Hirschman, E.C.: Differences in consumer purchase behavior by credit card payment system. J. Consum. Res. **6**(1), 58–66 (1979)
11. Holmström, J., Stalder, F.: Drifting technologies and multi-purpose networks: the case of the swedish cashcard. Inf. Org. **11**(3), 187–206 (2001)
12. Humphrey, D.B.: Retail payments: new contributions, empirical results, and unanswered questions. J. Bank. Financ. **34**(8), 1729–1737 (2010)
13. Humphrey, D.B., Kim, M., Vale, B.: Realizing the gains from electronic payments: costs, pricing, and payment choice. J. Money Credit Bank. **33**(2), 216–234 (2001)
14. Humphrey, D.B., Pulley, L.B., Vesala, J.M.: Cash, paper, and electronic payments: a cross-country analysis. J. Money, Credit Bank. **28**(4), 914–939 (1996)
15. Kahn, C.M., Roberds, W.: Why pay? an introduction to payments economics. J. Financ. Intermediation **18**(1), 1–23 (2009)
16. Knights, D., Noble, F., Vurdubakis, T., Willmott, H.: Electronic cash and the virtual marketplace: reflections on a revolution postponed. Organization **14**(6), 747 (2007)
17. Lawson, R., Todd, S.: Consumer preferences for payment methods: a segmentation analysis. Int. J. Bank Mark. **21**(2), 72–79 (2003)
18. Linné, T.: Digitala pengar. Nya villkor i det sociala livet. In: Lund Dissertations in Sociology, Lund University, Lund (2008)
19. Loke, Y.J.: Determinants of merchant participation in credit card payment schemes. Rev. Netw. Econ. **6**, 4 (2007)
20. Mallat, N.: Exploring consumer adoption of mobile payments: a qualitative study. J. Strateg. Inf. Syst. **16**(4), 413–432 (2007)
21. Mallat, N., Tuunainen, V.K.: Exploring merchant adoption of mobile payment systems: an empirical study. E-Service J. **6**(2), 24–57 (2008)
22. Penz, E., Meier-Pesti, K., Kirchler, E.: "It's practical, but no more controllable": social representations of the electronic purse in Austria. J. Econ. Psychol. **25**(6), 771–787 (2004)
23. Penz, E., Sinkovics, R.R.: Triangulating consumers' perceptions of payment systems by using social representations theory: a multi-method approach. J. Consum. Behav. **12**(4), 293–306 (2013)
24. Priem, R.L.: A consumer perspective on value creation. Acad. Manag. Rev. **32**(1), 219–235 (2007)
25. Raghubir, P.: An information processing review of the subjective value of money and prices. J. Mark. Res. **59**(10), 1053–1062 (2006)
26. Raghubir, P., Corfman, K.: When do price promotions affect pretrial brand evaluations? J. Mark. Res. **36**, 211–222 (1999)
27. Runnemark, E., Hedman, J., Xiao, X.: Do consumers pay more using debit cards than cash? Electron. Commer. Res. Appl. **14**(5), 285–291 (2015)
28. Schreft, S.L.: How And Why Do Consumers Choose Their Payment Methods?. DIANE Publishing, Collingdale (2006)
29. Simmel, G.: The Philosophy of Money. Psychology Press, Hove (2004)

30. Stavins, J.: Effect of consumer characteristics on the use of payment instruments. N. Engl. Econ. Rev. **3**, 19–30 (2001)
31. Xin, H., Techatassanasoontorn, A.A., Tan, F.B.: Antecedents of consumer trust in mobile payment adoption. J. Comput. Inf. Syst. **55**(4), 1–10 (2015)

Credit Scoring and the Creation of a Generic Predictive Model Using Countries' Similarities Based on European Values Study

Erika Matsak[✉]

Estonian Business School, Lauteri 3, 10114 Tallinn, Estonia
erikamatsak@gmail.com

Abstract. Starting with a new product in a new market brings companies a lot of risks and costs. There are companies, who can provide generic scoring models, but usually the accuracy of generic models is not sufficient and they are expensive. The possibility to create a generic predictive model based on a similar country model has been studied. The European Values Study and GESIS Data Archive have been used for research and the similarity coefficient has been calculated and used in the model. The results show that it is possible to build a new model using data from another, similar country and thus minimize costs and risks.

Keywords: Decision-making · Credit scoring · Similarity · Predictive modeling

1 Introduction

This paper introduces the possibility to create a generic predictive model based on a similar country model, which has been created before. Thanks to European Values Study and GESIS Data Archive it is possible to use the large spectrum of parameters which analyze people's data in all European countries. Many conclusions based on European research results have been published at the present moment[1]. The aim of this research is to find the possibilities to calculate the similarity coefficient using as much input as possible in order to create a predictive model based on this coefficient and to mathematically analyze and prove the benefits of this method in practice. At first the author introduces the aspects related to scoring and the problems related to start-ups in new markets. Next, the author uses Lorents metrics to analyze the four countries' similarities step by step and shows the possibilities to create a predictive model using the resultative coefficient and discusses the efficiency of the proposed approach.

---

[1] http://www.gesis.org/en/services/publications/gesis-papers/.

© Springer International Publishing AG 2017
S. Feuerriegel and D. Neumann (Eds.): FinanceCom 2016, LNBIP 276, pp. 114–123, 2017.
DOI: 10.1007/978-3-319-52764-2_9

2   Classification and Credit Scoring

A Credit scoring analyzes the user's personal information, which can be collected from different sources, starting from the authorization process on the credit company's web site, questionnaire data and any additional data collected from special bureaus, trusted systems etc.

After the client completes the questionnaire it is possible to use the parameters to create a classification algorithm. There are many different studies that suggest the methods for data collection and algorithms for classification [1]. From the theory of statistics we also know that this process starts from choosing the number of necessary classes for results. One way is to classify clients to two groups: good and bad clients. Sometimes the solution is related to more groups, for example we may also need to classify a medium class of clients. By choosing the classic statistical approaches, for example the logistic regression method, we can categorise the results to binary classes [2] or when using the multinomial logistic regression it is possible to categorise the results to a multiclass classification [3]. In credit scoring [4] we analyze the customer's information by using the chosen classification algorithm. This enables us to calculate the probability of the user belonging to a certain class (for example one class are good clients, the second class are bad clients). Logistic regression is the most often used method for credit scoring models and gives better classification results [5].

Unfortunately when starting business activities in a new country we usually don't have the necessary data for classification and that's why we have to ask ourselves: how can we predict the credit score of someone from a new country if the information required for classification is incomplete or is non-existent?

3   Start-up with a Product in a New Country

We will skip all the juridical aspects necessary for start-ups with a new product in a new country on the credit market; also we'll skip all the technical IT aspects for the authentication of clients and move directly to the creation of a scoring model.

The first important step in creating a scoring model is related to data selection. It is important to analyze the possibility to find third party credit check companies, who provide information related to credit depth (for example see Bisnode[2], Instantor[3]) and to understand what kind of information can be provided by them. The information provided by them will be reliable and parameters like debts, income, outgoings, other existing loans and credit cards, pending payments etc. will be verified.

If this is the first country and the first product for your company, you probably don't have a database with information about your customers, payments history etc. If you start from zero you can't create your own personal model. The solution in such a situation is to find the appropriate scoring company and outsource the scoring of clients. It is important to collect all the possible data provided during the scoring

---

[2] https://www.bisnode.com/international/.

[3] https://www.instantor.com/.

procedure and store it in your database. Scoring companies usually have several packages available and you can choose the appropriate one for your company.

In this paper we will discuss the possibility to use one country's previously created and well working model (Fig. 1) for a new product start-up in a new country. In other words we will use the model created for country C1 and predict the possibilities for country C2. Below is a list of questions:

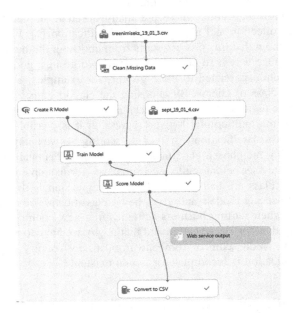

Fig. 1. Model for country C1 created in Microsoft Azure machine learning

- How to understand if the prognoses with input from country C2 and predicted in model C1 are successful and trustworthy?
- How to measure the similarities between the two countries?
- How to join the parameters with country-specific values (for example region from country C1 with region from country C2)?

For a moment let's forget the problems that different countries present. Let's move on to the logical-philosophical aspects. If we talk about two persons: A1 and A2, and presume that in situation S1 they will both prefer the same thing, what is their behaviour based on? A1 and A2 have sets of values and rules given to them as a child. Like in algebraic systems [6] we can separate a set of personal values (objects), a set of personal rules (relations), for example the importance of one value in a numerical measure and try to predict their behaviour. In mathematics we will present such an algebraic system as follows: $\langle$\{values\}, \{numerical measure\}, \{states\}; \{predicate of Importance\}, \{behavioural rules\}$\rangle$. A1 and A2 will behave identically if all sets in their algebraic systems are equivalent.

From psychology we know that behavioural rules have a strong relation to values [7] and thus we reach some states as a result of our behaviour. Therefore the most important role is played by values and the importance of these values.

This paper uses the European Values Study and GESIS Data Archive[4] [8]. *"The most comprehensive research project on human values in Europe. It is a large-scale, cross-national, and longitudinal survey research program on how Europeans think about family, work, religion, politics and society"*[5].

The last researches took place from 2008 to 2010 and discovered the moral, religious, societal, political, work, and family values of Europeans. According to the priorities in research and the developments in the current project, we will only compare four countries in this paper: Poland, Spain, Czech Republic and Sweden. The aim of this analysis is to understand how similar these countries are in their values and priorities and to predict how similar the people living in these countries are in their behaviour. The questionnaire has the same structure in each country and consists of more than 400 questions[6].

4 The Similarities in Countries' Values Methods

There are different methods available to analyse the questionnaire and to compare parameters. One of the best known methods is a chi-square test [9]. Using this method we will know how important the differences in the distribution of some parameters by statistical means are. Other well-known approaches are based on different metrics. For example Euclidean distance [10] which uses a simple calculation and gives the standard difference between two vectors.

In our research we have compared people's priorities when it comes to values. All answers have been coded to numbers and have been arranged separately for every country, these are then submitted to pairs $\langle Vi, Cj \rangle$, where V_i is the value with a higher priority than C_j. After that the Lorents distance [11] has been calculated. As a result four sets with ordered pairs are created (Poland, Spain, Czech Republic, and Sweden) and analyzed in comparison with the Poland set.

For the above mentioned analysis, an important preparation has been made. Only ordinal and numeric variables have been selected for the calculation using Lorents metrics. All columns with unsuitable values have been removed. Columns with values −5 (other missing), −4 (question not asked), −3 (not applicable) have been removed. After the removals, the number of columns (questions) has remained at 176.

The next problem was connected to the fact that different questions had a different number of possible answers. To overcome the problem, a special transition of answer codes to a scale of [0...10] was made. This is presented in numbers (Fig. 2, value label corresponding value). This transition is very important, because otherwise it would be

[4] EVS (2011): European Values Study 2008: Integrated Dataset (EVS 2008). GESIS Data Archive, Cologne. ZA4800 Data file Version 3.0.0, doi:10.4232/1.11004.

[5] http://www.gesis.org/en/services/data-analysis/survey-data/european-values-study/.

[6] https://dbk.gesis.org/EVS/Variables/compview.asp?db=QEVSLF&id=&add=ZA4800&var=&lang=&id2=&var2=&lang2=&vsearch=&vsearch2=&s1=1&s2=1&s3=1&bool=.

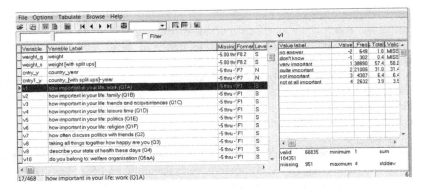

Fig. 2. Fragment from variable description

impossible to compare questions that have a different number of possible answers to choose from.

An explanation on how to move all the possible values to the same scale will be provided next.

Example: Let's take a look at two questions: v2 and v7. Question v2 – "how important is family in your life" can have the following answers: no answer (–2), don't know (–1), very important (1), quite important (2), not important (3), not important at all (4).

However question v7 – "how often do you discuss politics with friends" can be answered using another scale: no answer (–2), don't know (–1), frequently (1), occasionally (2), never (3).

At first the numeric value will shift to the positive scale:

- no answer (0), don't know (1), very important (2), quite important (3), not important (4), not important at all (5)
  - ● min = 0, max = 5
- no answer (0), don't know (1), frequently (2), occasionally (3), never (4)
  - ● min = 0, max = 4

Using the triangle rule (Fig. 3) and the corresponding formula, the following values will be calculated:

- no answer (0), don't know (2), very important (4), quite important (6), not important (8), not important at all (10)

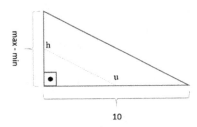

Fig. 3. Transformation to the same scale using triangle rule.

– no answer (0), don't know (2.5), frequently (5), occasionally (7.5), never (10)

As long as our variables are numerical or ordinal, it is possible to use the average calculation to compare the priority of the values. Table 1 shows a fragment of priorities by country, the variables are explained in Table 2.

Table 1. Priorities by country. Fragment.

SE	SE average	PL	PL average	CZ	CZ average	ES	ES average
v235	2.913	v235	3.092	v236	3.376	v239	2.851
v233	3.067	v236	3.349	v235	3.391	v235	2.882
v239	3.217	v239	3.365	v233	3.557	v238	3.372
v236	3.382	v238	3.512	v234	3.661	v244	3.508
v234	3.440	v244	3.651	v239	3.735	v234	3.601
v133	3.637	v249	3.657	v248	3.899	v236	3.671
v238	3.849	v248	3.705	v249	3.911	v250	3.740
v250	3.912	v234	3.785	v238	4.002	v233	3.770
v248	4.030	v240	3.880	v250	4.052	v252	3.930
v237	4.114	v233	3.925	v237	4.161	v247	4.048
v252	4.119	v133	3.992	v247	4.196	v133	4.067
v247	4.137	v246	3.998	v244	4.226	v249	1.4.03

Table 2. Explanation of variables of priorities. Fragment. For more information please see the official database (https://dbk.gesis.org/dbksearch/sdesc2.asp?no=4800ZA4800: European Values Study 2008: Integrated Dataset (EVS 2008))

do you justify:		
v235: joyriding	other missing	-5
v233: claiming state benefits	no answer	-2 (0)
v239: accepting a bribe	don't know	-1 (0.83)
v236: taking soft drugs	never	1 (2.5) (*definitely not)
v234: cheating on tax		2 (3.33)
v250: manipulation food		3 (4.17)
v248: prostitution		4 (5)
v237: lying in own interest		5 (5.83)
v252: death penalty		6 (6.67)
v247: avoiding fare public transport		7 (7.5)
v238: adultery		8 (8.33)
v133: do you believe that lucky		9 (9.17)
charm protects (*)	always	10 (10) (*definitely yes)

Measuring the priorities using Lorents metrics requires the values to be represented in pairs $\langle Vi, Cj \rangle$, where the value with more importance takes first place and the value with less importance is in second place [12]. For example using the fragment below (Table 1, see SE), it is possible to write out pairs $\langle v235, v233 \rangle$, $\langle v235, v239 \rangle$, $\langle v235, v236 \rangle$, $\langle v235, v234 \rangle$, $\langle v235, v133 \rangle$ etc.

The Matrix fragment of Spain and Poland (Fig. 4) illustrates which pairs Spain and Poland have. Here + indicates the pair's availability for Poland (values are placed to column headers and row headers) and * indicates the pair's availability for Spain.

	v1	v2	v3	v4	v5	v6	v7	v8	v9	v62	v63	v64	v65	v66	v92	v93	v94	v95	v96	v97	v98	v99	v100
v1		x+	x+	x+	x+	x+	x+	+		x+	x+	x	x+	x+		x+	+	x+	+	x+		x	x+
v2	x+		x+	x+	x+	x+	x+	x+	x+	x+	x+	x+	x+	x+	x+	x+	x+	+		x+	x+	x+	x+
v3				x+	x+	x+	x+	+	x+		x+	x+				x+		x+	+	x		x	x+
v4					x+	x+	x+	x+		x+			x+	x+		x+		x+	+	x		x	x+
v5								x+		x+													
v6				x+			x+			x+					+	+		+					
v7										x+													
v8				x+	x+	x+				x+			x+	x+		x+	+	+					
v9	x		x	x+	x+	x+	x+	x+		x+	x	x	x+	x+		x+		x+	+	x		x	x+
v62																							
v63				x+	x+	x+	x+	x+	+	x+			x+	x+		x+	+	x+	+	x+		x	x+
v64	+		x+	x+	x+	x+	x+	x+	+	x+	x+		x+	x+	+	x+	+	x+	+	x+	+	x+	x+
v65				x+	x	x+				x+						x+							
v66				x+	x	x+				x+													
v92	x+		x+	x+	x+	x+	x+	x+	x+	x+	x+	x	x+	x+		x+	x+	x+	+	x+	x+	x	x+
v93				x+	x+	x+				x+			x+	x+			+	+					
v94	x		x+	x+	x+	x+	x+	x+	x+	x+	x	x	x+	x+		x+		x+	+	x+		x	x+
v95				x+	x+	x+	x			x+			x+	x+	x		+	x					x
v96				+			+			+					+	+							
v97	+	+		x+	x+	x+	x+	+		x+			x+	x+		x+	+	+					x+
v98	x+		x+	x+	x+	x+	x+	x+	x+	x+	x+	x	x+	x+		x+	x+	x+	+	x+		x	x+

Fig. 4. Matrix of Spain (x) and Poland (+).

Relative differentiation for all values has been calculated by using the following formula $d(A, B) = [E(A) + E(B) - 2E(A \cap B)]:[E(A) + E(B) - E(A \cap B)]$, where $A \cap B$ is the intersection – or the pairs that Spain and Poland have in common, $E(A)$, $E(B)$ are the number of elements of all pairs in the corresponding countries and $E(A \cap B)$ is the element number of intersection.

The resulting similarity in values has been calculated as $1 - d(A, B)$ (Table 3). The results show that compared to Poland Sweden is more different than Spain and Czech Republic compared to Poland.

Table 3. Results

		d(A,B)	$1 - d(A,B)$
ES	PL	0.30778527	0.692215
CZ	PL	0.2929668	0.707033
SE	PL	0.40884435	0.591156

5   The Creation of a Predictive Model

The similarity coefficient can be used in model creation and calibration. First it shows whether or not countries can be joined to the generic model. Countries that are very different cannot efficiently be joined in the predictive model. In addition, the similarity coefficient can be used in a separate column as a predictive variable; it can also help analyze the probability that the client is a "good client". Logistic regression shows the connection between the logarithm (*logit*) and the linear combination of predictive variables, calculated using coefficients (b_o ... b_k) (Eq. 1): where p_i is the probability that the i-customer is "bad", x_i is the value of an independent predictive variable and ε is a random error component.

$$\ln\left(\frac{p_i^*}{1 - p_i^*}\right) = \ln\left(\frac{p_1 \pi_0}{p_0 \pi_1}\right) + b_0 + b_1 x_i^{(1)} + b_2 x_i^{(2)} + \ldots + b_k x_i^{(k)} + \varepsilon_i \tag{1}$$

If the similarity coefficient has been used as value xi, and we have a group of clients for model verification, where we know the real clients' behaviour then we can tune the bi coefficient to achieve a more accurate predictive result. The calculation of the probability pi using logistic regression is shown in Eq. 2.

$$p_i = \frac{1}{1 + exp(-(b_0 + b_1 x_i^{(1)} + b_2 x_i^{(2)} + \ldots + b_k x_i^{(k)} + \varepsilon_i))} \tag{2}$$

6   Discussion

When joining different countries to a common model it is sometimes important to join the variables that have the same parameter meaning, for example region or a third party score, however the values are different in each country. Here it is possible to use different types of mapping between the values by using the economic values from statistical databases or use special mathematical transitions from one scale to another. To improve the model, machine learning algorithms must be used in order to retrain the model, adding new country data to the previously trained dataset.

The first month of monitoring of the generic model created for Sweden which was based on Polish data has shown quite good results (Fig. 5). ROC curve which was used to test the 110 rows of Swedish data, made it possible to find out if the client paid the loan, or if the loan went to debt collection. As long as retraining is used after every 100 rows of data, we can expect better results in the future.

It is important to mention that some credit scoring companies, which can be used for scoring purposes, only provided a 0.7 accuracy on the ROC curve. The traditional academic point system is:

- 0.9–1 = excellent
- 0.8–0.9 = good
- 0.7–0.8 = fair
- 0.6–0.7 = poor
- 0.5–0.6 = fail

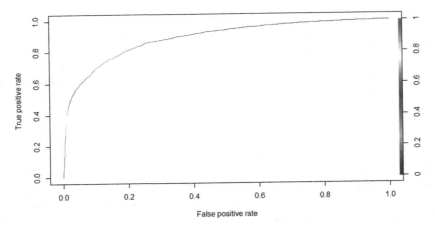

Fig. 5. ROC curve for Sweden.

7   Conclusions

Using a similarity coefficient to create a generic predictive model can help financial companies to start business activities in a new market. The similarity is calculated using Lorents metrics which is based on a strong mathematical background. The author of this paper has implemented the created model in one private financial company who has started activities in the Swedish market and a few months of monitoring have shown that the company has received profit. Poland and Sweden are not as similar as for example Poland and The Czech Republic; the author believes that the same approach can help the mentioned finance company to get profit in Czech and Spanish markets as well.

Future works could be related to model calibration using the similarity coefficients. The author hopes to find a better algorithm to connect the parameters, which need more attention. For example the correlation of education and salary in Sweden and in Poland is different and if the measure of correlation will be calculated with similarity coefficients, then it can help connect these parameters more efficiently.

This approach enables finance companies to invest with lower risks to the new markets immediately after the start-up.

The author would like to thank the finance company, where the author had the opportunity to test the novel ideas and colleague Taavi Rekor for the English language support. Also the author thanks professor Peeter Lorents for his assistance in preparing this work.

References

1. Sadatrasoul, S.M., Gholamian, M.R., Siami, M., Hajimohammadi, Z.: Credit scoring in banks and financial institutions via data mining techniques: a literature review. J. AI Data Min. **1**(2), 119–129 (2013)
2. Hauser, R.P., Booth, D.: Predicting bankruptcy with robust logistic regression. J. Data Sci. **9**(4), 565–584 (2011)
3. Li, T., Zhang, C., Ogihara, M.: A comparative study of feature selection and multiclass classification methods for tissue classification based on gene expression. Bioinformatics **20**(15), 2429–2437 (2004)
4. Sorokin, A., Сорокин А.: Developing scorecards using logistical regression. Построение скоринговых карт с использованием модели логистической регрессии/Интернет-журнал\"Науковедение\", Вып. 2 (21) (2014)
5. Gouvêa, M.A., Gonçalves, E.B.: Credit risk analysis applying logistic regression, neural networks and genetic algorithms models. In: POMS 18th Annual Conference (2007)
6. Maltsev, A.I., Мальцев, А.И.: Algebraic systems. Алгебраические системы. —Наука‖. Москва (1970)
7. Bardi, A., Schwartz, S.H.: Values and behavior: strength and structure of relations. Pers. Soc. Psychol. Bull. **29**(10), 1207–1220 (2003)
8. EVS: European Values Study 2008: Integrated Dataset (EVS 2008). GESIS Data Archive, Cologne. ZA4800 Data file Version 3.0.0 (2011). doi:10.4232/1.11004
9. Yates, F.: Contingency table involving small numbers and the χ^2 test. Suppl. J. R. Stat. Soc. **1**(2), 217–235 (1934). JSTOR 2983604
10. Deza, E., Deza, M.M.: Dictionary of Distances. Elsevier, Amsterdam (2009)
11. Lorents, P., Lorents, D.: Applying difference metrics of finite sets in diagnostics of systems with human participation. In: Proceedings of the International Conference on Artificial Intelligence, Las Vegas, USA, vol. III, pp. 1297–1301. CSREA Press (2002)
12. Alas, R., Lorents, P., Übius, Ü., Matsak, E.: Corporate social responsibility: European and Asian countries comparison. In: Toim. Sarinastiti, N., Lengkong, F., Sintawati, M. (eds.) Entrepreneurship in Global Competition, pp. 481–498. Penerbit Universitas Atma Yala, Jakarta (2012)

Author Index

Printed in the United States
By Bookmasters